Computational Methods for Numerical Analysis with R

CHAPMAN & HALL/CRC
Numerical Analysis and Scientific Computing

Aims and scope:
Scientific computing and numerical analysis provide invaluable tools for the sciences and engineering. This series aims to capture new developments and summarize state-of-the-art methods over the whole spectrum of these fields. It will include a broad range of textbooks, monographs, and handbooks. Volumes in theory, including discretisation techniques, numerical algorithms, multiscale techniques, parallel and distributed algorithms, as well as applications of these methods in multi-disciplinary fields, are welcome. The inclusion of concrete real-world examples is highly encouraged. This series is meant to appeal to students and researchers in mathematics, engineering, and computational science.

Proposals for the series should be submitted to one of the series editors above or directly to:
CRC Press, Taylor & Francis Group
3 Park Square, Milton Park
Abingdon, Oxfordshire OX14 4RN
UK

Published Titles

Computational Methods for Numerical Analysis with R

James P. Howard, II

The Johns Hopkins University
Applied Physics Laboratory
Laurel, Maryland, USA

CRC Press
Taylor & Francis Group
Boca Raton London New York

CRC Press is an imprint of the
Taylor & Francis Group, an **informa** business

A CHAPMAN & HALL BOOK

CRC Press
Taylor & Francis Group
6000 Broken Sound Parkway NW, Suite 300
Boca Raton, FL 33487-2742

First issued in paperback 2020

ISBN-13: 978-1-4987-2363-3 (hbk)
ISBN-13: 978-0-367-65791-8 (pbk)

Visit the Taylor & Francis Web site at
http://www.taylorandfrancis.com

and the CRC Press Web site at
http://www.crcpress.com

*To Nina, Chase (see Figure 4.13),
and Ducky (see Figure 4.14)*

Contents

List of Figures

List of Tables

List of R Functions

Preface

Computational Methods for Numerical Analysis with R(CMNA) is a treatment of the traditional numerical analysis course using R as the underlying programming language. The traditional numerical analysis outline begins with numerical error, then linear algebra, interpolation, integration, optimization, and differential equations. In this way, the outline covers the entire introductory mathematically sequence. This text is suitable for the advanced undergraduate student or first year graduate student. The book will require a solid understanding of linear algebra, differential and integral calculus, and differential equations. Students of mathematics, computer science, physics, engineering, and other mathematically intensive disciplines will have sufficient background to read and understand the book.

My motivation for writing this book is the lack of similar materials in the market. There are several commonly used textbooks that teach numerical analysis using MATLAB®. Others use C or Fortran. MATLAB® is a very expensive program and while available on college campuses, is not cheap enough for most graduates to purchase. C and Fortran are inexpensive or free, but require basic programming skills to manage data input and output, memory, and other tasks that should not be set upon someone trying to learn a specific set of skills. R provides a rich environment that students are already familiar with due to its rapidly growing user base. It is free for all users and it does not require intensive "environmental management" when programming, as is required in, for instance, Java.

The primary audience for this book are individuals interested in a deeper understanding of the methodological approaches used in numerical analysis, especially as they apply to computational statistics and data science. The audience should have a solid grounding in mathematics through calculus and differential equations. Data scientists can also use this text as the starting point for implementing new analytical techniques in R.

This book presents a different approach to computational methods. Rather than focusing on theory and proofs, we focus on the value of working code, and shy away from formal proofs. The code is demonstrated, as functioning, directly in the text. Using working code as our guide, our goal is to reach an intuitive understanding of the process each algorithm uses through explanation, code, and examples. With that understanding, we can modify, support, and improve these codes. More importantly, we can understand what R, or some other language, is doing when it executes its internal functions to solve numerical problems.

Each section of this text present a type of problem and a numerical solution. The problem is described briefly and the numerical solution is explained. Then the R code implementing that solution is described, along with any specific implementation details. From there examples are given. Where appropriate, examples showing where the algorithm can fail are provided along with suggestions for where robust implementations are available as part of the standard R distributions, or one of the many packages available from CRAN.

Because this book focuses on working code, this book is not meant to be a textbook. However, its structure mirrors the structure of an undergraduate computational methods course. As such, this text would be appropriate an textbook for courses entitled Numerical Analysis, Computational Methods, Scientific Computing, or similar courses, if the instructor is willing to forego rigours analytical approaches. CMNA would better serve as a secondary text for one of these courses when the text provides weaker examples, or in an inaccessible manner, such as Fortran or MATLAB®.

I would like to thank the following Rob Calver, Lara Spieker, and many others at CRC Press for seeing this book through to completion. I would also like to thank Marck Vaisman, Elizabeth Shope, Kelley O'Neal, James Long, and Casey Driscoll, who all offered great comments at one point or another. I also want to thank two anonymous reviewers for extensive and insightful comments on the text. Any errors remaining are only my responsibility.

1

Introduction to Numerical Analysis

1.1 Numerical Analysis

Numerical analysis is the branch of mathematics that addresses solving mathematical problems with computers in a search for numerical answers, as opposed to symbolic answers. These numerical solutions provide answers suitable for practical and real-world applications. In contrast, symbolic solutions are exact, but may be difficult or impossible to find and offer more detail than can practically be applied.

R was developed to provide a free implementation of the S programming environment for statistics. Like S, R can provide solutions to statistical analysis similar to STATA and SAS. R also provides a comprehensive programming environment and a number of tools for processing functions and matrices. With these tools, R provides an environment for calculation and can be used to solve numerical problems. This text will provide a guide for doing numerical analysis in R.

1.1.1 The Goals of Numerical Analysis

Numerical analysis is an applied branch of mathematics. It seeks out solutions to practical problems that can be put into practice. The practical aspect of it leads to a desire for solutions that are "good enough." But good enough is harder to define. The value and suitability of numerical solutions are constrained by three concepts. First, we are concerned with the efficiency, or how much work is necessary to reach a solution. Second, we are concerned with the accuracy of our solution, or how close we are to the true value. Finally, we are concerned with precision, or the level of detail in our solution. We will examine each of these concepts in turn and learn how to apply them to numerical analysis. But good enough is entirely dependent on the application space.

Numerical analysis underlies many fields which require numerical answers to mathematical questions. From video games to climate science, and physics to psychology, numerical analysis performs the calculations we use daily. Engineers need to know the tolerances for load bearing capacities for a building. Financiers need to calculate the expected market return on an investment.

And medical professionals use numerical analysis to process advanced imagery. Numerical analysis supports almost any field when computers are used to solve mathematical problems.

Numerical analysis is different from other branches of mathematics. In calculus, we expend considerable effort to identify the exact value of equations, even if they cannot be represented by a "number" such as 7, 21, or 98.6, because the value is irrational or transcendental. Accordingly, numerical analysis is the engine of applied mathematics.

For instance, we know a circle's area is πr^2 and the area of a circle that is 20 meters across has a radius of 10 meters and is 100π square meters. But for determining how much mulch is necessary for a 10-meter-across circular garden, we really need an actual number of square meters. When it comes to this sort of practical problem, knowing there are about 314 square meters is probably good enough.

This distinction is what separates numerical analysis from symbolic computation. Symbolic computation would result in the solution of 100π. There are platforms for providing these symbolic computations. Both Mathematica by Wolfram Research and Maplesoft's Maple packages provide symbolic computation engines. Free alternatives in both Maxima and SAGE are also available and all of these products are capable of providing numerical results, as well.

But in a symbolic computation platform, expressions are first class objects, like numbers are to R. Except for some minor cases around linear models, R does not work well with symbolic computation on expressions. For instance, adding x to $x + y$ should yield $2x + y$, a concept R cannot handle. However, if x and y have values associated with them, R will dutifully find the sum of $2x$ and y. Despite this limitation, the D function family will symbolically find derivatives of some mathematical expressions.

However, the constraints listed above affect the final results, especially when the constraints work at cross purposes. Accordingly, numerical analysis is not just the search for numerical solutions to mathematical problems. It is also the search for usable estimates of those solutions. These usable estimates might be bounded, so while we might know if they are correct, we might know the error is no more than some limit. From there, we can evaluate how usable our estimate is.

Applied mathematics is sometimes considered as distinct from pure mathematics. However, applications of mathematics are based on the underlying pure mathematics. Often, numerical methods are straightforward implementations of the related pure mathematical solution for the problem. Because of this relationship, some problems are better understood after applying pure mathematics. Accordingly, when a function generates an error, it can sometimes be resolved through symbolic manipulation.

In addition to the three constraints, numerical analysis is subject to several important limits. The most important is on size of the data. Modern computers have vast amounts of memory. But for some problems, no amount of memory is sufficient. And for others, no amount of computational time is sufficient. These space and time limits are continually pushed back as computers grow faster and larger. Today's complex climate modelling was a dream ten years ago and a fantasy ten years before that. But even straightforward problems, such as the determinant of a matrix, using the traditional method of calculation, grow unchecked, quickly.

Another limitation is the availability of good algorithms for solving numerical problems. For instance, in Section 1.3.2, we will review three different algorithms for evaluating polynomials. These have different qualities and different benefits. These limitations are baked into the process in a way the constraints are not. We can relax a constraint. In fact, we can usually trade one constraint for another. But the limitations are externally imposed. We can build bigger and faster computers, but only so much and so quickly. Further, we can evade some limitations through the use of parallel programming, not addressed in this text. Parallel programming allows processes to be broken into multiple independent steps and executed, simultaneously, on different computers with the final results aggregated at the end.

1.1.2 Numerical Analysis in R

R was developed based on the S programming language to provide a domain-specific language for statistics. R, like S, grew out of the Unix environment and provides a modular environment that is extensible and flexible. R has developed a reputation for implementing new algorithms quickly, delivered through a package management system that makes installation of new modules easy.

R is also free software. Like many free software packages, R is developed primarily by volunteers or those working for companies using the software. Those volunteers are often the authors of new modules. This has led to a strong support community through mailing lists and websites dedicated to R.

Even though R is geared toward and known primarily as a statistical analysis environment, the platform provides a complete programming language with numerical primitives. Providing a complete programming environment, along with attendant editors and other supporting applications, makes it possible to implement any algorithm in R. The language includes specific features for using and managing datasets and numerical computation. Because of its power and ready availability, R is a good environment for learning numerical analysis.

R's rise as a statistical analysis platform has introduced the language and its fundamentals to a wide audience. If you are already familiar with R from a statistical perspective, you are familiar enough with the language for numerical analysis. Many software packages exist to provide numerical analysis. However, the large complete packages, especially MATLAB®, are very expensive and may cost more than many students and independent learners can afford. Finally,

```
fibonacci <- function(n) {
    if(n == 0)
        return(0)
    if(n == 1)
        return(1)
    return(fibonacci(n - 1) + fibonacci(n - 2))
}
```

R Function 1.1
The Fibonacci sequence

because R provides a complete programming environment, any algorithm can be implemented in it.

In this text, implementations of algorithms are presented in set off boxes and numbered. For instance, Function 1.1 includes an implementation of the Fibonacci number generator. Examples of R input and output are given set off within the text, lending the example to the narrative, like this,

```
> fibonacci(10)
[1] 55
> 3 + 8
[1] 11
```

showing the tenth member of the Fibonacci sequence and the results of three plus eight. Input and output are shown as fixed-width text and input lines begin with a bracket (>). The examples are directly run in R and added to the LATEX source for this text using Sweave. Accordingly, they show the actual output from R without cutting and pasting or other potential corruptions. The version used is

```
> print(version$version.string)
[1] "R version 3.3.2 (2016-10-31)"
```

In developing this text, a few rules were adopted. First, a function must be well formatted. This means functions have whitespace, usually empty lines, to break up internal components. Lines must be short enough not to wrap, either. However, control structures, like if statements or for loops are set off through indentation, a common approach in formal coding. If a line from a function becomes too long, the line would be rewritten or shortened to fit, horizontally, across the page. This could involve using other approaches such as splitting a complicated operation into more than one step. A final strategy is that internal variable names are not always fully descriptive. For instance, instead of a variable called *upperBoundaryLimit* or some other lengthy noun phrase, something shorter such as *upper* or even *u* may be used. Variables should be unambigious and easy to remember in context, regardless.

In addition, R code examples must be small enough to fit on a single printed page in this text. Most are under half a page. This encourages the reader to open the text and read functions. The small size is designed to help example implementations fit in the head of the reader. As a result, error handling and potential evaluation options are not included, unless convenient or necessary. In some cases, smaller independent components were moved into new functions. This provides additional flexibility to the implementation and improves the readability of the functions.

The functions included in this book, and some others, are bundled into a package called `cmna`, standing for *Computational Methods for Numerical Analysis*. This package provides the algorithms, implementations, documentation, and testing. The package is available through standard distribution channels.

```
> library(cmna)
```

The call to `library` makes the library available for use. Despite bundling these functions into a single package, the functions provided do not provide a unified whole. That is, the functions provide a baseline for understanding the imeplementation and are designed to provide a comprehensive solution for numerical applications. Where possible, functions with similar applications are given similar interfaces. This is not to say the functions were not designed. But they are not designed to work together in any meaningful way.

Because of these design choices, it is also important to remember the functions implemented here are not robust. That is, they do not perform basic error handling and they are unlikely to fail gracefully when presented with a problem. This is the tradeoff from the simple implementations given. They explain the fundamental method from a numerical standpoint. Handling unusual circumstances or errors, should be a routine part of any program, but because they are not included here, some implementations may not fail well. As such, error handling is addressed when necessary, such as when a function could silently return an incorrect answer, rather than simply terminate.

The functions in `cmna` are not robust and should not be used for actual applied numerical analysis. Applied numerical analysis, such as an architectural model, relies on well-tested and well-understood numerical models that handle errors. R includes robust and tested functions for numerical analysis. Because the functions we develop are not robust enough for everyday use, when appropriate, these functions are described and demonstrated to show how to solve actual numerical analysis problems in R. This allows you to move along and actually use the demonstrated solution rather than fixing the `cmna` version to meet more demanding requirements.

1.1.3 Efficiency

The first constraint we will look at is efficiency. Programmers generally work very hard to make sure their computers do not have to. Every gain in efficiency enables more interesting problems to be solved. Because programs for numerical analysis are run so many times, increased efficiency decreases the overall runtime, as well. Efficiency comprises two areas. The first is time complexity, focusing on how long it takes for a program to execute. Programs that require more steps for larger inputs are said to have higher time complexity. The second is space complexity, measuring how much storage, even intermediate, is necessary to execute the program. Programs that require more memory for larger inputs are said to have higher space complexity.[1] We are going to use less formal methods focusing on the count of operations a function must perform to complete and return a value. For instance, examine this loop:

```
> count <- 0
> m <- 10
> for(i in 1:m)
+      count <- count + 1
> count
[1] 10
```

This is a fancy way to perform a loop 10 times. The function loops from 1 to m times. This loop's internals execute 10 times. To verify this, we increment the *count* variable and print the value, which is m. Of course, we can also nest loops, and this increases the number of times the loop's internals, the parts within the innermost loop, run:

```
> count <- 0
> m <- 10
> n <- 7
> for(i in 1:m)
+     for(j in 1:n)
+          count <- count + 1
> count
[1] 70
```

In this case, we have two loops. The outer loop runs from 1 to m and the inner loop executes from 1 to n. The inner loop executes n times for each time through the outer loop. Therefore, the loop's innermost internals execute mn times. These are straightforward examples. Slightly less straightforward is an example where the outer loop executes from 1 to m. The inner loop executes from 1 to i where i represents the iteration of the outer loop:

```
> count <- 0
> m <- 10
> for(i in 1:m)
+     for(j in 1:i)
+          count <- count + 1
> count
[1] 55
```

---

[1]For a textbook treatment, I recommend *Introduction to Algorithms* by Cormen et al. (2009) or *Algorithms* by Sedgewick and Wayne (2011). For a less formal approach, an Internet search for "time complexity," "space complexity," or "big-$\mathcal{O}$ notation" is sufficient.

```
isPrime <- function(n) {
    if(n == 2)
        return(TRUE)

    for(i in 2:sqrt(n))
        if(n %% i == 0)
            return(FALSE)

    return(TRUE)
}
```

R Function 1.2
Test an integer for primality

The inner loop's internals execute 1 time the first time through the outer loop, 2 times the second time, and so on up to m times the mth time through the loop. Accordingly, the inner loop executes k times where $k = \sum_{i=1}^{m} i$.

Each of these grow at least as quickly as the input does. However, that is not always the case for a function. Consider the function isPrime, listed in Function 1.2. This function performs a basic primality test by testing the input for division by a successive number of potential divisors.

When testing for primality, it is only necessary to test the integers from 2 to the last integer less than or equal to $\sqrt{n}$. This function still grows, but at a much slower rate. Further, the relative rate of increase decreases as n increases. Also, in the case of isPrime, $\sqrt{n}$ is the maximum number of times through the loop. If the number is not prime, an integer less than $\sqrt{n}$ will successfully divide n and the loop will terminate early.

Of course, within a loop, it is possible for complex operations to be less obvious. In this example,

```
> count <- 0
> m <- 10
> x <- 1:100
> y <- rep(1, 100)
> for(i in 1:m) {
+     y <- y * x
+     count <- count + 1
+ }
> count
[1] 10
```

the value of *count* is 10. However, many more multiplications happened within the loop. The loop itself contains an element-wise multiplication. This is a multiplication for each element of the vectors x and y. As each are 100 elements long, this loop actually contains 1000 multiplications. Despite there being only one `for` loop within the example as stated, it includes more operations than the last example, with its explictly nested loops.

Understanding what is happening inside a loop is critical to estimating how long it will take to execute. If the loop calls other external functions, each of those will have their operations to perform. Likely, these functions are opaque to us, and we will not necessarily know what they are doing and how they do it.

Despite these speed analyses, R is not a fast programming language. Its internal operations are fast enough for most purposes. If speed were the most important consideration, we would use C or Fortan. As compiled languages, they execute as machine code directly on the hardware. R programs, like Python, MATLAB®, and other interpreted programming languages, do not run directly on the hardware, but are instead translated and executed by the language's runtime engine.

But we use languages like R, MATLAB®, or Python for other reasons and in the case of R, the reasons might be access to libraries created by other programmers, the statistical facilities, or its accessibility. And understanding how to estimate the number of operations necessary to complete a task is language-neutral. So if one implemented a C version of the prime tester in Function 1.2, it would still require $\sqrt{n}$ tests.

1.2   Data Types in R

R provides a wide variety of data types both natively and through extensions. These data types include atomic elements, which are elements that consist of a single element of data. These are Boolean data types, those that are true or false, and numbers. More complex datatypes exist that are treated as core objects by R. This includes dates, strings, matrices, and arrays. Finally, R supports a number of compound data types that include multiple elements such as the results of a regression.

We will be focusing on a few data types that directly support numerical analysis, either as the input or the output. These data types are the purely numerical types, such as floating point numbers. These data types also include aggregate types from linear algebra, vectors, and matrices. Finally, our discussion will include data frames, a special type of matrix with named properties that is optimized for use by many of the statistical functions in R. First, we will begin with Boolean data types, which provide critical support to all of R.

1.2.1   Data Types

R presents a number of ways to store data. Many of these are variants on storing numbers. The fundamental data types are building blocks for compound and aggregate data types. The first data type is for Boolean values. These can take on two values, either TRUE or FALSE, are intended to represent their logical counterparts. These values, returned from some functions, are often passed to functions to set a flag that affects the operations of the function. Like many programming languages, R can treat these Boolean values like they are numbers. The Boolean value of TRUE is converted to 1 and the Boolean value of FALSE is converted to 0:

```
> TRUE == -1; TRUE == 0; TRUE == 1; TRUE == 3.14
[1] FALSE
[1] FALSE
[1] TRUE
[1] FALSE
> FALSE == -1; FALSE == 0; FALSE == 1; FALSE == 3.14
[1] FALSE
[1] TRUE
[1] FALSE
[1] FALSE
```

This is different from many programming languages, however. With R, all values other than 0 (FALSE) and 1 (TRUE) are not equal to either Boolean value. In addition, TRUE and FALSE are ordered such that TRUE is greater than FALSE.

```
> TRUE < FALSE; TRUE > FALSE
[1] FALSE
[1] TRUE
```

In addition, these Boolean values can be operated on by many functions within R. For instance, because TRUE can be treated as 1, you can use sum function to count the number of TRUE values in a vector. Other functions work as expected, too. For instance, the following example creates a vector of five Boolean values of which three are true. In addition to the function sum, the function mean and function length return the expected values:

```
> x <- c(TRUE, FALSE, TRUE, FALSE, TRUE)
> sum(x); mean(x); length(x)
[1] 3
[1] 0.6
[1] 5
```

In a similar way, we can use this to count the number of entries in an array that are equal to a certain number:

```
> x <- c(1, 2, 3, 4, 3, 1, 2, 3, 3, 4, 1, 3, 4, 3)
> sum(x == 3)
[1] 6
```

Or, extending the model, we can count how many entries are less than 3:

```
> sum(x < 3)
```

```
[1] 5
```

R provides two principal data types for storing numbers. The first of these are integers. In R, integers are simple integer data types. These data types do not provide for any decimal component of a number. They are whole numbers and are treated as such. There are no numbers in between these integers and the distance between each integer is 1 unit. The mechanics and implications of this are further explored in Section 2.2.1. While having no fractional component to a number may seem like an unnecessary hindrance, the data type is available for use when special circumstances encourage it. For instance, we might use an integer to represent the year, as the fractional component can be represented separately, as a month and day count, which are also integers.

For those applications which require a fractional component to the number, there is also the numeric data type. The numeric data type is the base data type. It is the default data type when new numerical data is entered and most computations return numeric types. It is stored as a floating point number within the computer and the mechanics and implications of that are explored in Section 2.2.2. Floating point numbers are used to represent any number with a fractional component, such as a distance measurement or simply, $1/2$.

R provides built-in functions we can use to test the type of a variable and to convert among data types. We sometimes call the conversion process coercion. First, the function `is.numeric` will return true if the argument given is numeric and the function `is.integer` will return true if the argument is an integer. Both functions return false, otherwise:

```
> x <- 3.14
> is.numeric(x)
[1] TRUE
> is.integer(x)
[1] FALSE
```

The `is.numeric` and `is.integer` functions test the internal data storage type, not the actual data, so testing 3 for integer status will be false, as a raw number, when entered, is numeric by default:

```
> is.integer(3)
[1] FALSE
```

Similar test functions exist for Boolean and many other data types within R. Conversion is almost equally simple. The functions `as.numeric` and `as.integer` will convert a number to the named data type. At the most extreme, these conversion features will extract numerical data from strings. Suppose the variable x is a string containing the "3.14" and we wish to treat it as a number:

```
> x <- "3.14"
> as.numeric(x)
[1] 3.14
> as.integer(x)
```

`[1] 3`

The `as.integer` function converts any number to an integer. This is done by truncating the fractional part, which is equivalent to rounding down. If the argument to the function is, for instance, 3.9, the conversion to an integer will still yield 3. Data type conversion is not the best method to use for rounding fractional numbers R. There is a powerful `round` function that is designed to address the place selection and other aspects of rounding rules.

Like the family of functions for data type testing, there is a family of conversion functions that can convert among many different data types within R. For instance, we can convert a numerical value to a string using the function `as.character`, which converts a number to its decimal representation, suitable for display.

Integer and numeric data can usually be freely mixed and matched. R will, unless otherwise directed, convert the internal representations into whatever form is most appropriate for the data being represented. This is almost always going to be the numeric data type, and for almost any purpose, that is the best option for us, as well. The numeric data type is versatile enough and capable enough to handle our common data storage needs. In addition, as R's principal data type, any function we will want to use will understand it implicitly.

In addition to these data types, R supports a number of nonnumerical data types. Strings are shown in the last example. R supports two data types called factors and ordered factors, which are simply label types. The labels can be considered as categories which are explicitly unordered. The ordered data types are factors with an inherent ordering. These are frequently used in R to support statistical applications. R also supports a number of date formats and other specialized aggregate types. Generally, these data types are not often used in purely numerical problems, but may be the end result of an applied problem.

1.2.2 Data Structures

Using its broad collection of numerical and nonnumerical data types as a base, R is extended with a collection of compound and aggregate data structures. These data structures support the variety of applications R has been put to use for, such as times and dates for time series, or images for image processing. However, two of these data structures will be useful throughout this text. The vector and the matrix are simple homogenous aggregate data structures, in one- and two-dimensional formats. These underlie many different applications in numerical analysis.

The vector object represents a single dimensional ordered list of values. This corresponds to the standard linear algebra notion of a vector, and we will use them as such. Vectors can be created using the internal R function c, which collects the values and orders them, like this,

```
> (x <- c(1, 0, 1, 0))
[1] 1 0 1 0
```

which creates a simple vector of alternating 1s and 0s. This is the same nota-tion we used previously to create an array of Boolean values for an example. We can use the c function to add additional values to the vector and to append one vector to another:

```
> (y <- c(x, 2, 4, 6))
[1] 1 0 1 0 2 4 6
> (z <- c(x, y))
 [1] 1 0 1 0 1 0 1 0 2 4 6
```

An individual element of the vector can be addressed using square brackets (the brackets that look like [ ]) to enclose the element number:

```
> z[10]
[1] 4
> z[c(10, 9, 1)]
[1] 4 2 1
```

In the second example, a vector of values 10, 9, and 1 is the referenced address of the object, z, to get the tenth, ninth, and first elements of the vector, respectively. Importantly, the elements are returned in the order specified.

List data types are similar in appearance to vectors, but have different applications in practice. Lists are created in a similar fashion, but using the list function in R. Lists are more versatile. For instance, while all elements of a vector must have the same data type, lists can aggregate different data types, such as numbers and strings, into a single object. In addition, list items can be named, making lists the aggregation container for complex data elements. The following example creates a list called $z1$ with element a equal to 3 and element b equal to 4 and a list called $z2$ with element s a string and *nine* equal to 9.

```
> (z1 <- list(a = 3, b = 4))
$a
[1] 3

$b
[1] 4
> (z2 <- list(s = "test", nine = 9))
$s
[1] "test"

$nine
[1] 9
```

In addition, lists are recursive. That is, we can embed one list within another, whereas vectors concatenate.

```
> (z <- list(z1, z2))
[[1]]
[[1]]$a
[1] 3

[[1]]$b
[1] 4


[[2]]
[[2]]$s
[1] "test"

[[2]]$nine
[1] 9
```

We will use lists for some return values, when it is important to return a result in more than one part. But otherwise, we will not use lists for mathematical purposes.

The second important aggregate data type is the matrix. Matrices are the heart of linear algebra, And because linear algebra underlies much of numerical analysis, as we shall see, the matrix is the core data type we will use. Inside of R and other numerical analysis platforms, very fast linear algebra routines accomplish many of the calculations we rely on.

There are several ways to create a matrix in R. The first we will look at is the matrix function.

```
> (A <- matrix(1:12, 3, 4))
     [,1] [,2] [,3] [,4]
[1,]    1    4    7   10
[2,]    2    5    8   11
[3,]    3    6    9   12
```

The matrix function takes three principal options. The first is a list of data elements to populate the matrix. The second is the number of rows and the third is the number of columns. The number of elements should equal the number of rows times the number of columns. If the number of columns is omitted, the matrix function will assume a sensible value, if possible, such as the total number elements divided by the number of rows given.

As shown in the display of matrix A above, each column and row is numbered. We can address individual rows and columns of a matrix using those numbers. We can also address individual elements by specifying both the row and column. The syntax is similar as for a vector:

```
> A[2,3]
[1] 8
> A[2,]
```

```
[1]  2  5  8 11
> A[,3]
[1] 7 8 9
```

Note the label for the row includes a number, a comma, and nothing else. In contrast, to specify a column begins with a comma, and the column number. In each case, there is an empty specification where the comma separates nothing from the addressed row or column and the returned type for both is a vector. This syntax also matches the column and row labels shown in the displayed matrix A. Finally, specifying both the row and column yields a single element.

In addition to assembling a matrix using the `matrix` function, it is possible to create a matrix from vectors. Two functions, `rbind` and `cbind` will append vectors as rows and columns, respectively, to existing matrices. If the existing matrix is instead another vector, the two vectors shall become a matrix:

```
> x1 <- 1:3
> x2 <- 4:6
> cbind(x1, x2)
     x1 x2
[1,]  1  4
[2,]  2  5
[3,]  3  6
> rbind(x1, x2)
     [,1] [,2] [,3]
x1    1    2    3
x2    4    5    6
```

Note in the `cbind` example, each column is named, and in the `rbind` example, each row is named. In both cases, each is named after the source vector. However, all can be addressed through numerical addressing.

Another important operation on matrices is transposition. Transposition mirrors a matrix across the main diagonal. To transpose a matrix in R, use the `t` function. Using the example of A above:

```
> (A)
     [,1] [,2] [,3] [,4]
[1,]   1    4    7   10
[2,]   2    5    8   11
[3,]   3    6    9   12
> t(A)
     [,1] [,2] [,3]
[1,]   1    2    3
[2,]   4    5    6
[3,]   7    8    9
[4,]  10   11   12
```

Of course, we can use the `t` function on the output to return the original matrix.

The R function t will also transpose a vector. That is, it will convert a "horizontal" vector into a "vertical" vector and vice versa. However, the method is not necessarily intuitive because a vertical vector is really stored as a single-column matrix. Using the t on a vector will first return a matrix with 1 row and a column for each element vector. Then a second call to the t function will transpose the new matrix into a vertical vector: Using the example variable x, from above:

```
> t(x)
     [,1] [,2] [,3] [,4]
[1,]    1    0    1    0
> t(t(x))
     [,1]
[1,]    1
[2,]    0
[3,]    1
[4,]    0
```

It may be clear from these examples, but vectors and matrices, in R, are closely related. In linear algebra, the vector is silently treated as a single-dimensional matrix, in either orientation. This simplifies the rules for multiplying matrices by vectors and this is suitable for most applications. However, in R, a vector is not a single-dimensional matrix. A vector is a distinct object that can be coerced into a single-dimensional matrix. Frequently, however, a vector can be treated as a single-dimensional matrix, provided element addressing is not used. If element addressing is being used on a vector, it can be advantageous to convert it to a single-dimensional matrix for clarity.

In addition to vectors and matrices, there is another important aggregate data type in R, the data frame. Usually, a matrix is expected to have the same data type for each element. In our examples, every element has been numeric. A data frame will usually have many different data types, such as Boolean objects, timestamps, or even strings. Though implemented internally as specialized lists, data frames are two-dimensional objects, like a matrix, without the expectation of homogenous data types.

Like lists, we may use data frames to store return values conveniently, but will not often use them otherwise. Data frames can be created with the data.frame function, which works similarly to the matrix function. In addition, the cbind and rbind functions will work with data frames in the expected way.

Finally, aggregate data structures present a different question relating to missing data. Data could be missing for a variety of reasons, such as an observation was not possible or was lost. In some cases, data may be removed intentionally for different purposes. In the simplest case, we can simply declare the data was not available.

R provides a special value, NA, that represents missing data. Testing the value NA against any other value, including itself, yields a result of NA. Because of this propagating NA, R provides a function to test for NA values, the `is.na` function.

```
> NA == NA
[1] NA
> NA == 1
[1] NA
> natest <- c(1, 2, NA, 4, 5)
> is.na(natest)
[1] FALSE FALSE  TRUE FALSE FALSE
```

We might reasonably infer the missing value is 3, simply by guessing. However, R does not know what value belongs there and uses the missing value placeholder.

Using vectors and matrices, R provides all of the tools necessary to implement linear algebra. With linear algebra, R can also provide a full suite for numerical analysis. As we move forward, we will introduce more concepts about vectors and matrices as we need them. In particular, Section 3.1 will provide much more information about working with vectors and matrices, including basic linear algebra operations.

1.3   Elementary Problems

With this introduction in mind, we now move to some basic algorithms. These are designed to illustrate the diversity of approaches we can use even with the most elementary problems. For instance, we will present three different algorithms to sum a vector of numbers. Each of these examples will also help introduce basic concepts about error analysis and efficiency, as we formally define the concepts.

In addition to summation, we will discuss methods for evaluating a polynomial, and use that method later in the text. However, the polynomial methods show how basic algebra can be used to create more efficient and better implementations. Finally, we will show an iterative approach to solving the nth root problem. Iterative approaches will be used throughout this text to solve many different problems and for the nth root, we can see its distilled form.

1.3.1   Summation Algorithms

Computers are good at highly repetitive tasks. A basic task that we want to perform often is summing a vector. Mathematically, this is the simple expression,

$$S = \sum_{i=1}^{n} x_i. \tag{1.1}$$

```
naivesum <- function(x) {
    s <- 0
    n <- length(x)

    for(i in 1:n)
        s <- s + x[i]
    return(s)
}
```

R Function 1.3
The naïve approach to summing a vector

But implementation may not be so simple. Equation 1.1 is critical in other contexts, most notably in finding variable means and standard deviations. Finding ways to ensure the correctness and efficiency of the vector sum is important across many different applications.

The straightforward method for finding the vector sum, we call the naïve approach, cycles over each entry in the vector and adds it to a cumulative sum. The cumulative sum is initialized to 0 and each value of the vector is added in sequence. The naïve approach executes one addition for each member of the vector, as each value is added to the running sum. An R implementation of the straightforward method, here called `naivesum`, is provided in Function 1.3. The function returns a numerical value of the total sum:

```
> x <- c(1, 2, 3, 4.5, -6)
> naivesum(x)
[1] 4.5
```

R's looping is very slow compared to some other programming languages. Among other reasons looping can be slow is the test to exit the loop. In this implementation, the test is implied in the `for`-loop, when the code determines if the index i has exceeded the maximum allowed value, n. In general, we will discuss more abstract aspects of the implementation and their efficiency.

Of course, because addition for rational numbers obeys the associative property, it does not matter what order the numbers are added in. For instance, the sum could be calculated in reverse order. Or any order. We will assume there is a function $f()$ such that,

$$f(x_1, \ldots, x_n) = \begin{cases} x_1 & : n = 1, \\ f(x_1, \ldots, x_{n/2}) + f(x_{n/2+1}, \ldots, x_n) & : \text{otherwise}. \end{cases}$$

The function $f()$ is recursive, meaning it calls itself. The function works by continuously splitting its input in half and reinvoking itself on each half separately. If the function has reached the point where there is only one member of the input vector, it simply returns the value of the single member. When the result is returned to the previous iteration of the pairwise function, it will be

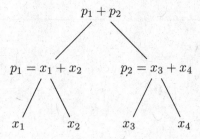

Figure 1.1
Assembly of partial sums in piecewise addition

```
pwisesum <- function(x) {
    n <- length(x)

    if(n == 1)
        return(x)
    m = floor(n / 2)
    return(pwisesum(x[1:m]) + pwisesum(x[(m + 1):n]))
}
```

R Function 1.4
The pairwise summation algorithm

added to the result of another call to the pairwise function, creating a partial sum. The partial sums filter up through the tree of recursive function calls and the final result is the total of all these partial sums, each added as pairs. This is shown in Figure 1.1.

We call this the piecewise summation algorithm. This function works because of the associate property of addition and the relative simplicity of the overall operation. A complete implementation of a piecewise summation algorithm is presented in Function 1.4. From the user's viewpoint, the `pwisesum` function acts just like the `naivesum` function. Assuming the value of x is unchanged:

```
> pwisesum(x)
[1] 4.5
```

The actual implementation given in Function 1.4 makes some allowances for the potential real-world implementation problems. Most importantly, it is unlikely the number of elements in the vector x will be power of 2. As a result, the division of labor between the two uses the `floor` function to establish the integers below and above the one-half demarcation point for finding the split. In practice, this has no effect on the implementation nor the execution.

```
kahansum <- function(x) {
    comp <- s <- 0
    n <- length(x)

    for(i in 1:n) {
        y <- x[i] - comp
        t <- x[i] + s
        comp <- (t - s) - y
        s <- t
    }
    return(s)
}
```

R Function 1.5
The Kahan summation algorithm

The piecewise summation algorithm uses almost as many internal additions to compute the final sum. In fact, where the naïve approach resulted in $n-1$ additions, the piecewise summation algorithm also uses $n-1$ additions, in the best case. We will assume we have a vector of eight numbers and we wish to find the sum, S,

$$
\begin{aligned}
S = 1 &+ 2 + 3 + 4 + 5 + 6 + 7 + 8 \\
3 &\quad + \quad 9 \quad + \quad 11 \quad + \quad 15 \\
&12 \quad\quad + \quad\quad 16 \\
&\quad\quad 28.
\end{aligned}
\tag{1.2}
$$

Equation 1.2 shows the steps in a pairwise addition of eight numbers. For each plus sign $(+)$ shown on the four lines comprising S, there is an addition; this example has seven. This is the same number necessary to add the numbers sequentially using the naïve approach.

Also interesting is the number of times the function is called. Normally, a nonrecursive function will only be called once, for each instance. A recursive function is called many times. In this case, there is a call for each addition in the process. This is represented by the plus signs, above. Because of the way the function is implemented, it is called once for each single member vector. Other implementations may choose to add the elements of a vector, if there are only two, and return that value. This is normally an implementation detail left to the user creating the function.

We can also choose to process the vector in a completely different way. The Kahan summation algorithm works in one such way. The function is given in R as `kahansum` in Function 1.5. This function is developed as a method to compensate for numerical error, a concept we will formally introduce in the next chapter. Here, it is presented as another alternative approach to the problem of summation.

The function captures a small piece of the total for later use. This value, called the compensation (represented by the variable *comp*) is initially set to 0, along with the running total, the variable s. After being set to 0, the function begins looping over the elements in the input vector. For each element of the loop, four steps are executed. First, a temporary variable, y, is the difference between the current x element and the compensation. Then a second temporary value, t, holds the sum of the current x element and s. At this point, the compensation is recalculated and so is the running sum.

Mathematically, the running sum should be the actual running sum and the compensation should always be 0 because the computer should execute the sums perfectly. In practice, this may not be the case. Regardless, the kahansum function finds the sum as expected:

```
> kahansum(x)
[1] 4.5
```

Each of these example functions demonstrates different approaches to calculating a vector sum. These approaches each have positives and negatives. If the straightforward approach is sufficient to meet the operational goals, its simplicity may well outweigh any other concerns about speed.

Because R is dominated by and created for statisticians, the needs of statisticians come first. Statisticians use functions like the standard deviation and mean. These require summation algorithms to implement. R includes the sum function in the base platform. The sum function is robust and provides advantages over our homegrown solutions. For instance, the sum function has an option to eliminate the values of NA from the summation:

```
> sum(c(1, 2, 3, 4, NA, 5))
[1] NA
> sum(c(1, 2, 3, 4, NA, 5), na.rm = TRUE)
[1] 15
```

Despite these niceties, sum itself is implemented using a sequential addition algorithm, just like naivesum. However, the internal R function is implemented in C and should be substantially faster than our pure R implementation.

1.3.2   Evaluating Polynomials

Like summation, polynomial evaluation is a problem which allows for multiple solutions. Unlike summation, there are large differences in the evaluation of the naïve approach versus other solutions. We will begin by defining our polynomial. Let $f(x)$ be a polynomial such that,

$$f(x) = a_n x^n + a_{n-1} x^{x-1} + \cdots + a_1 x + a_0. \tag{1.3}$$

Every polynomial has an identical form and structure. There is a collection of coefficients attached to a value of x with its exponent. There is little difference, conceptually, between $f(x) = x^2 - 1$ and $f(x) = x^9 - 3x^2 + x - 99$.

```
naivepoly <- function(x, coefs) {
    y <- rep(0, length(x))

    for(i in 1:length(coefs)) {
        y <- y + coefs[i] * (x ^ (i - 1))
    }

    return(y)
}
```

R Function 1.6
The algebraic approach to evaluating polynomials

Accordingly, a convenient form of representing polynomials can be useful. A vector can be used to "store" a polynomial because it will contain all of the information necessary to replicate that polynomial. This occurs in two ways. First, each coefficient can be stored as a value in the vector. Second, degree of the polynomial, which equals the total number of coefficients, is the length of the vector. Provided we remember that the final coefficient, without an x term and a_0 in equation 1.3 is equivalent to $a_o x^0$, this method works perfectly.

We will use the vector to store them and each coefficient will be an entry in the vector, with the lowest order polynomials appearing first. That is, the first entry in the vector is the coefficient a_0, the second is a_1, and so forth. For the polynomial $f(x) = x^2 - 1$, the vector is `c(-1, 0, 1)`. Having developed a convention for polynomial notation and storage, we will now show how to evaluate them. For example, consider the polynomial,

$$f(x) = (x - 5)(x + 3)(x + 2)(x - 1)$$
$$= x^4 + 3x^3 - 15x^2 - 19x + 30. \tag{1.4}$$

We will represent this polynomial as a vector and use it for examples in this subsection:

```
> f <- c(30, -19, -15, 3, 1)
```

In the naïve approach, we will evaluate the polynomial in roughly the same manner as we would in a high school algebra class. We will substitute the value of x into the function, exponentiate it as appropriate, and then multiply by the coefficient. Each term is then summed and the value is returned. The naïve approach is implemented in the R in the function **naivepoly**, which is given as Function 1.6.

The implementation opens by creating a vector, y, with the same length as the length of the input value, x, and setting each element to 0. This allows the **naivepoly** function to accept and return evaluations of a polynomial at a vector of x locations.

Then the function loops over each element of the coefficient vector, from

```
betterpoly <- function(x, coefs) {
    y <- rep(0, length(x))
    cached.x <- 1

    for(i in 1:length(coefs)) {
        y <- y + coefs[i] * cached.x
        cached.x <- cached.x * x
    }

    return(y)
}
```

R Function 1.7
A better algebraic approach to polynomial evaluation

the first to the nth, in that order. For each time through the loop, which is the variable i in the function, the algorithm executes a single addition and i multiplications. The i multiplications are composed of $i - 1$ multiplications in the exponentiation and one more for the coefficient. For a polynomial of degree n, that is n additions and $\sum_{i=0}^{n} i = (n^2 + n)/2$ multiplications. The number of multiplications grows at the rate of the square of the degree of the polynomial.

For the function $f(x)$ constructed before, evaluation at a point x reveals its simple use. Consider evaluating $f(-1)$, $f(0)$, and $f(1)$ using the `naivepoly` function:

```
> x <- c(-1, 0, 1)
> naivepoly(x, f)
[1] 32 30  0
```

We can increase the efficiency of polynomial evaluation. For instance, each time through the loop, we find the exponentiation of x. But for each coefficient, a_i, the associated value of the exponentiation, x^i, is the product series,

$$x^i = \prod_{j=0}^{i} x. \tag{1.5}$$

For x^i, there are i instances of x multiplied together. But for x^{i-1}, there is one fewer x and so on down to $x^0 = 1$. We can use this by caching the value of each multiplication and storing it for the next iteration.

Function 1.7, `betterpoly`, implements this. It uses a variable *cached.x* to keep the running value of the product series available for the next iteration of the loop. Compared to the function `naivepoly`, there are two multiplications and a single addition for every time through the loop. For a polynomial of degree n, there are $2n$ multiplications.

This is a substantial improvement over `naivepoly`. For a polynomial of

```
horner <- function(x, coefs) {
    y <- rep(0, length(x))

    for(i in length(coefs):1)
        y <- coefs[i] + x * y

    return(y)
}
```

R Function 1.8
Horner's rule for polynomial evaluation

degree 10, `naivepoly` executes 55 multiplications and `betterpoly` executes only 20 multiplications. The number of multiplications in `naivepoly` could be reduced by one if the calculation of *cached.x* were eliminated for the final execution of the loop.

There are other options for calculating polynomials. Starting with equation 1.3, we can rewrite a polynomial such that,

$$
\begin{aligned}
f(x) &= a_n x^n + a_{n-1} x^{x-1} + \cdots + a_1 x + a_0 \\
&= a_0 + a_1 x + \cdots + a_{n-1} x^{x-1} + a_n x^n \\
&= a_0 + x(a_1 + \cdots + a_{n-1} x^{x-2} + a_n x^{n-1}) \\
&= a_0 + x(a_1 + \cdots + x(a_{n-1} + x(a_n)) \cdots).
\end{aligned} \tag{1.6}
$$

Equation 1.6 algebraically reduces a polynomial to n multiplications and n additions. This is an increase in efficiency over `betterpoly`. For a polynomial of degree 10, it is only 10 multiplications. This algorithm is called Horner's method and is named for William Horner, who discovered the methods in the early 18th century.

An implementation of Horner's method is given in Function 1.8, `horner`. This implementation unwinds the summation of equation 1.6 and turns it into an R loop. For each iteration of the loop, there is one addition and one multiplication. Reducing the number of multiplications reduces the amount of time necessary to evaluate a polynomial at a given value of x.

A modern computer is so fast that it will not matter if you only execute the evaluation a few times. However, some applications, such as root finding or numerical integration could execute a function thousands of times. Accordingly, we benefit from faster execution times, even for a simple polynomial function which otherwise seems trivial. Speed, in this instance, does not sacrifice the quality of the output. Both `betterpoly` and `horner` provide the same result as the naïve method:

```
> betterpoly(x, f)
[1] 32 30  0
> horner(x, f)
```

`[1] 32 30  0`

Like the implementation of `betterpoly`, `horner` includes an unnecessary step. On the first iteration of the loop, x is multiplied by the partially complete sum, which had been initialized to. But similar to other cases, the extra step is permitted here to simplify the overall implementation. In addition, if a conditional if-then test were included, it would be executed on every iteration of the loop, likely at greater computational expense than the single unnecessary multiplication.

These types of code simplification steps usually require some sort of analysis like this to determine if they improve the program's execution and readability. In some cases, the gains of readability may be far outweighed by any speed gains that could be made that reduce simplicity. Many factors go into these decisions including future code reuse and the business requirements of the operating environment.

There is no built-in function or function set that provides polynomial evaluation. However, the `polynom` package includes functions for polynomials and polynomial division. The `polynom` package includes dedicated objects for storing polynomials, including specialized access functions. The functions included here do not create a class for managing internal data. Accordingly, the functions in the `polynom` package are more robust than the implementation provided here.

1.3.3   The nth Root Algorithm

Finally, we look at the nth root algorithm. This algorithm is unlike the others presented so far because it is iterative. That means the algorithm executes a loop and after each iteration of the loop, the result becomes progressively better. The loop can be terminated when the result is good enough for the application. This process is known as convergence.

The algorithm works by setting the initial value, x_0, to some guess at the value of the nth root of a. The successive values of x are,

$$x_{i+1} = \frac{1}{n}\left[x_i(n-1) + \frac{a}{x_i^{n-1}}\right]. \tag{1.7}$$

This is a special case of Newton's general root finding algorithm, which we will discuss more in Section 6.1.2. For now it is best to assume the algorithm is correct.

$$\Delta x_i = x_{i+1} - x_i = \frac{1}{n}\left[\frac{a}{x_i^{n-1}} + x_i(n-1)\right] - x_i$$

$$= \frac{1}{n}\left[\frac{a}{x_i^{n-1}} + nx_i - x_i - nx_i\right]$$

$$= \frac{1}{n}\left[\frac{a}{x_i^{n-1}} - x_i\right]. \tag{1.8}$$

```
nthroot <- function(a, n, tol = 1 / 1000) {
    x <- 1
    deltax <- tol * 10

    while(abs(deltax) > tol) {
        deltax <- (1 / n) * (a / x ^ (n - 1) - x)
        x <- x + deltax
    }

    return(x)
}
```

R Function 1.9
The *n*th root algorithm

The algorithm, implemented as **nthroot** and presented in Function 1.9, begins by setting the value of *x* to an initial guess at the root. For this implementation, we set the initial guess to 1, regardless of the value of *a* or *n*. Unless *a* itself is 1, an initial guess of 1 is guaranteed to be incorrect. However, almost any guess will do, except for 0, since that will cause division by zero in Equation 1.8.

The value of *deltax*, representing Δx, is initially set to 10 times the value of *tol*, the maximum tolerance in the solution. In the main loop, *deltax* is calculated and added to *x*, once for each pass through the loop. The loop continues until such time as *deltax* is smaller than the *tol*. At that point, the value of *x* will be no further from the true value of *x* than *tol*. Formally, assuming the true value of the *n*th root is x^*, then $|x - x^*| \leq \epsilon_t$, where ϵ_t is the tolerance. Following the loop, the returned value of *x* includes the estimate:

```
> nthroot(100, 2)
[1] 10
> nthroot(65536, 4)
[1] 16
> nthroot(1000, 3)
[1] 10
> nthroot(pi, 2)
[1] 1.772454
```

The algorithm converges rapidly. However, R includes a built-in solution for finding the *n*th root that is more reliable, but it is not a function. The ^ operator for exponentiation accepts decimal exponents. Recall from algebra that $\sqrt{x} = x^{\frac{1}{2}}$ and, generally, $\sqrt[n]{x} = x^{\frac{1}{n}}$. Therefore, the standard approach for finding the *n*th root of a number in R is to take the number to power of $1/n$:

```
> 100^(1/2)
[1] 10
```

```
> 65536^(1/4)
[1] 16
> 1000^(1/3)
[1] 10
> pi^(.5)
[1] 1.772454
```

Accessing the objects as a primitive operator provides a substantial benefit to a tolerance-limited solution encouraged by the nth root algorithm.

Comments

From here, we have established some basic information and introduced numerical analysis. We have distinguished what we do in numerical analysis from symbolic computing, the realm of both computer algebra systems and pen and paper. And we have also introduced the basic data types and structure that provide R with the power and elegance necessary to do numerical analysis. With a basic toolkit, we finally took a look at a number of algorithms to solve common problems in mathematics.

Using these tools, we are going to explore a number of problems that arise in applied mathematics across fields. Before we get to more algorithms, we need to talk about error and what it means for the computer to be wrong. The computer can be wrong for a lot of reasons, and some of them we can limit. In other cases, there is little we can do other than recognize the problem and brace ourselves.

The algorithms for division and summation range from straightforward to very complex, and for good reason. Each has a place and an application. Sometimes an algorithm for a numerical problem will be very fast, but only apply in a subset of cases. Or it may produce a result that is not as accurate. Numerical analysis then moves from mathematics to an art, as we select the algorithms and tools necessary to meet our needs for the least amount of computing effort. Sometimes meeting this requirement, however, requires more human and programming effort.

For instance, review the `isPrime` function in Function 1.2. In this function, given a number n, we test every integer between 2 and $\sqrt{n}$ inclusive in the search for divisors. However, this process is painfully more intense than need be. For instance, after only a moment of thinking, we realize that if 2 is not a divisor of n, then no multiple of 2 will be a divisor n, either. Accordingly, we should probably not check any event number except 2. And while that alone would halve the number of tests to complete the evaluation, it is only a partial help. After all, the same logic extends that if 3 is not a divisor, neither is any multiple of 3, and with 5, and 7, and so on. A better algorithm than presented would remove unnecessary and repetitive evaluations.

These sorts of analysis are the core of numerical analysis. We want to find the best possible algorithm to suit the task at hand, but the task goes beyond the problem as stated. It includes how fast a solution must be calculated, how accurate it must be, and how much time we have to implement a solution. All of these come together as our constraints when solving numerical problems and each has a different aspect. Project managers often say you can have a solution fast, cheap, or correct, provided you pick only two. The same sort of constraints play out here.

The rest of this text will develop an understanding of error analysis, and the meaning of both precision and accuracy. Then we will develop a set of tools for performing tasks in linear algebra, followed by a suite of tools for interpolation, of various types. Then we will move into calculus and discuss algorithms for both differentiation and integration, optimization, and finally a review of methods for differential equations.

Exercises

1. What is the circumference of a 10-meter-diameter circle symbolically? What is that to 3 decimal places? Which one is correct?

2. Describe an application for which a numerical solution were preferable to a symbolic solution. Describe an application where symbolic solutions are preferable to numerical.

3. If an unsigned variable were stored using 16 bits, what is the largest number it could hold?

4. Using R's internal trigonometric functions, implement a secant function.

5. Using the exponentiation operator, create a function that finds the nth root of any number.

6. Function 1.1 provides a recursive implementation of the Fibonacci sequence. Rewrite the function to provide a result without recursion. What are the relative strengths and weaknesses of each implementation?

7. Implement a product function in R without using recursion.

8. Implement a product function in R using recursion.

9. Find the Horner polynomial expansion of the Fibonacci polynomial, $F_6(x) = {}^5 + 4x^3 + 3x$.

10. Reimplement Function 1.7 to eliminate the calculation of *cached.x* on the final iteration of the loop.

11. The Horner polynomial evaluation method in Function 1.8 is implemented as an iterative loop over the coefficient terms. Reimplement the Horner method as a recursive function. List the advantages and disadvantages of this method.

12. Reimplement Function 2.3 or 2.4 to allow for negative dividends and divisors.

13. Using equation 1.8, find the fifth root of 16807.

14. When painting a large room, determine the degree of precision necessary for how much paint to purchase.

15. Evaluate each of the polynomial function calculators and determine the best use case for each.

16. Both data frames and lists support named elements. Describe a case to use a compound data type and whether it is better stored as a list or a data frame.

17. Use R to determine if `TRUE` and `FALSE` are ordered. What implications does this have?

18. Reimplement the piecewise summation algorithm without using recursion. Does this implementation have any specific applications?

19. Develop a new algorithm for testing primality that does not test multiples of already checked divisors.

2

Error Analysis

2.1 True Values

Rounding numbers is common in everyday conversation. We do this because getting an answer that is close enough is sufficient for most purposes. For instance, we might give directions and say "your destination is around five miles past the old post office," assuming you know which post office is which. Or we might say something costs twenty dollars when the actual price is $21.99. This extends beyond purely numerical values. Al Jolson had an important date at "about a quarter to nine," and it is unlikely his date arrived at precisely 8:45 p.m.

Numerical analysis is grounded on getting answers that are good enough. A calculation should target the highest precision and accuracy necessary without wasting computational resources. This is because numerical analysis finds approximations for everyday use in science and engineering. A modern 3D printer may have a printing resolution of 100 microns (a micron is one-one millionth of a meter and 100 microns is one-tenth of a millimeter). When constructing the computer model of something to be printed, describing the curve with any finer resolution than the printer may accept is unnecessary. Further, it wastes computing resources.

Conversely, the numbers fed into an algorithm for evaluation may be only close approximations. The distance to the sun is commonly given at 150 million kilometers. But the Earth's orbit is an ellipse and the distance is constantly fluctuating. Similarly, a meter stick may not have any marks closer together than one millimeter. An observer may estimate where between two marks a measurement lies, or may pick one of the endpoints.

In addition, the rule for selecting an endpoint can follow different rounding rules. For instance, the observer may be required to pick the smaller of the potential options. Or the observer may be required to pick the larger of the potential options. Finally, an observer may select whichever mark appears closest. These rules may be complicated if the rule for observational rounding is unknown or perhaps not strictly enforced.

The level of correctness necessary to get a final answer is dependent upon both the input numbers and the final application's requirements. The amount of error we can allow in a calculation is dependent entirely on the circumstances. We call this the error tolerance. The error tolerance for cooking can

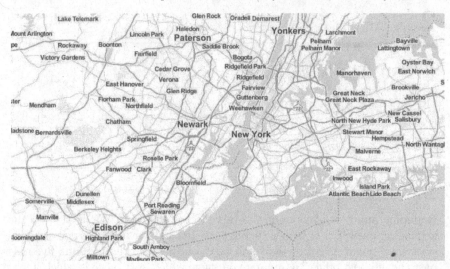

Figure 2.1
Newark is near New York (Stamen Design)

be quite high. When making cookies, measurements for flour or sugar can be "about a cup" or even undefined. It is unclear just how much salt is in a pinch. On the other hand, when working with chocolate, temperatures have to be controlled very carefully or the chocolate may not crystallize correctly when cooled.

Numerical error is composed of two components, precision and accuracy. Precision and accuracy are two related terms that are often used interchangeably. These terms describe two different facets of how an estimated value relates to the true value. But each has a different meaning and a different application in numerical analysis.

For this section, we will assume there is a true value, x, that we are attempting to measure. We will further assume that x' is the measured value. The value of x' may be measured through observation or may be the result of a numerical process or calculation. Finally, the error will be represented by ϵ.

2.1.1 Accuracy

Accuracy measures how close an estimate is to the true value. This is the commonly used definition for accuracy. For example, if we were travelling to New York City, we might get directions. If we get directions to Newark, New Jersey, we will end our trip in the wrong city, the wrong state, and on the wrong side of the Hudson River. This is a problem of accuracy. Directions to Boston or London are progressively less accurate than our directions to Newark. But sometimes close is close enough, depending on how the estimate could be used. New York is a brief train ride from Newark. How much accuracy we need depends on the application the estimate is used for.

We can measure accuracy in different ways. The first is absolute accuracy or *absolute error*. This measures the raw distance between the correct value of x and the measured value. Formally,

$$\epsilon_A = |x - x'|, \tag{2.1}$$

where ϵ_A is the absolute error. This measurement is a raw figure and, importantly, in units of the underlying measure. If x is in kilometers, so is x'. Accordingly, the value of ϵ_A will also be denominated in kilometers and we can say our estimate was off by ϵ_A kilometers.

However, the absolute error of a measurement only partially conveys the context of the measurement and may not present a meaningful error estimate in context. In the directions, Newark is 16 kilometers (10 miles) from New York. That is a substantial difference in this context. But in another context, a 16-kilometer error may not make a substantial difference. For instance, using the radius of the Earth's orbit around the Sun again, an error of 16 kilometers is not very large. This is especially true considering the Earth's orbit varies from 147 million to 151 million kilometers from the Sun.

Measuring how substantial the estimated error is leads to a different error type definition. The *relative error* measures, as a part of the true value, how much error there is in the measurement. Formally,

$$\epsilon_R = \left| \frac{x - x'}{x} \right|, \tag{2.2}$$

where ϵ_R is the relative error. By conveying the error as a proportion of the true value, the effect of the error can be judged and evaluated.

Unlike the absolute error, the relative error is unitless. It is the ratio of the error to the true value and the units cancel each other. This has advantages. The unitless error estimate can be compared across different systems, if otherwise appropriate. Or the relative error can be interpreted as a percentage of the true value. Reducing relative error has a larger impact on correctness than reducing the absolute error, when considered as a percentage of the true value.

However, even with a small relative error, it is possible for the effect to cause problems. Being in Newark when a job interview is in New York is catastrophic. In contrast, after crossing hundreds of millions of kilometers to go to another planet, if a space probe were off by 16 kilometers, it might crash into the planet, or it just might miss the best picture opportunity. With each estimate of the accuracy of measurement, the context is a guide to how suitable the accuracy is.

Accuracy only tells the story of how far from the true value a result is. There are limits on how close the true result can be. If a sequence of digits is infinite, as the digits of π are, then no computer contains sufficient memory

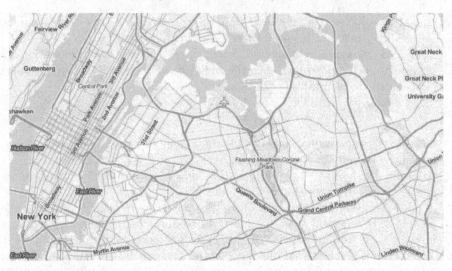

Figure 2.2
At least we got to New York, this time (Stamen Design)

and computational power to calculate a result using the full number of digits. Practically, this is rarely a concern and the limits of the calculation are described as the precision of the result.

2.1.2   Precision

Precision provides a measure for how specific a number is. That is, precision explains the level of detail in a measurement. Continuing our travel analogy, getting directions to New York when we are interested in going to the Empire State Building is a problem of precision. We have gone to the right place, but the right place is too large to get to the final destination. Directions to Manhattan narrow the directions and you might see the Empire State Building. But directions to the corner of 34th Street and Fifth Avenue will take you to the building itself. As the directions increase in precision, we get closer to our destination, without being incorrect at each step.

It is possible to estimate a value of x with an apparent high degree of precision, but this may not necessarily be more accurate. For instance, we can estimate the value of π at 22/7, a commonly used approximation,

$$\pi \approx \frac{22}{7} \approx 3.14285714\ldots \tag{2.3}$$

This estimate is accurate to two decimal places. But the estimate given by 22/7 is a decimal that continues forever. Only the number of digits we are willing to show limits its precision. However, the value of this estimate is its accuracy and for some purposes, this level of accuracy is sufficient.

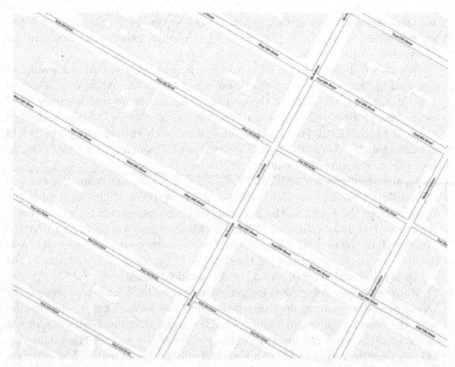

Figure 2.3
A journey of a thousand miles starts with a good map (Stamen Design)

A more accurate estimate of π would simply be the first 8 digits, 3.14159265. This estimate, while less precise that 22/7 has greater accuracy and is usable in many applications. Finding the area of a circle from the radius ($A = \pi r^2$) will be more accurate with this simple representation than with 22/7. However, the area in both cases can be calculated using any number of digits. This is known as *multiple precision*. While multiple precision mathematics libraries are widely available, they are mostly implemented in computer algebra systems. R is a system designed for computation and most calculations in R have limited precision.

While increased precision may sound beneficial, it can cause problems. A calculation may induce greater precision in the final result than is warranted. This can cause users of those calculations to trust them more than they should. This is called *false precision*. False precision arises when a very precise estimate for a measure is presented, but the precision presented does not necessarily reflect the precision actually included in the number.

For instance, a note saying a box will hold 5 kilograms might turn into 11.0231 pounds. In this case, the precision is generated by the conversion factor, 2.20462 pounds per kilogram. But to say the box will hold 11.0231 pounds adds a measure of precision that is probably unwarranted, and certainly easy

to misunderstand. The critical input value was the 5-kilogram figure. It has
one significant digit, the 5. The final result should include only one, and be
interpreted as 10 pounds.

Significant digits, also called *significant figures*, are the parts of a number
that include the precision of that number. This is limited to any non-zero
digits in the representation of the number. The significant digits convey all of
the precision of a number. For instance, the number π can be represented as
3, 3.1, 3.14, 3.142, 3.1416, 3.14159, and so on with an increasing number of
significant digits. Accordingly, as the number of significant digits increases, so
does the precision of the estimate.

Significant digits are frequently encoded in scientific notation. The number
1.5×10^8 is the same as 150 million and 150,000,000. When discussing the
distance from the Earth to the Sun, this estimate was shown to be the center
of a range. But as an estimate, 150 million kilometers is a good fit and 1.5×10^8
conveys all of the critical information. The 1.5 carries all of the precision and
the 10^8 carries the magnitude of the number. As a result, this estimate of the
radius of the Earth's orbit contains two significant digits.

However, when doing mathematics with numbers of limited precision, the
lowest precision number determines the final precision. That is, the number
with the fewest significant digits included carries that number of significant
digits into the final calculation. That is why in the case of the 5-kilogram
holding box, the estimate in pounds should only be 10, no matter how precise a
conversion factor is used. Critically, this statement holds for multiplication and
division of estimated values, but also for addition and subtraction. However,
unlike with addition and subtraction, multiplication and division can change
the magnitude without affecting the number of significant digits in the final
result.

Coming back to rounding, the process of rounding numbers reduces both
the number of significant digits and the precision of an estimated value. For
each place of rounding, the number of significant digits is reduced by one.
With π, rounding to 6 total digits is 3.14159, because the next number is 2.
However, if the value of π is rounded to only 5 digits, the value is 3.1416.
In this case, a significant digit is lost and a single digit of precision is lost.
Accuracy is also lost, but by an unknown and bounded amount. In this case,
the rounding process happened in the last place and split the input range
evenly. As a result, the upper bound on the absolute error is 0.00005. We
know the error is actually smaller than that, but only because we know the
value of π.

Because a computer only has a finite amount of storage, we can only store
numbers with a limited degree of accuracy and precision. In practice, there are
standard representations of numbers that limit the accuracy and precision to
a fixed value. Two important representations are the integer and the floating
point number.

2.2 Internal Data Storage

Internally, R has a variety of ways to store numbers. We have already seen the aggregate data types such as matrices and vectors. However, the two storage types we are going to focus on here are integer and numeric data types. Most numbers in R are stored as floating point numbers. These are numbers that allow for storing decimals. We will begin our discussion of data storage with integers.

R does not use integers often and even when used as an index for a vector, integers are usually stored as floating point numbers. As we will see, floating point numbers are made up of several components, including two integers that tell us the value and size of the number. So understanding how integers are stored and used gives us a better understanding of how floating numbers work.

2.2.1 Binary Numbers

When R creates an integer, as we described in Section 1.2.1, R uses the native 32-bit integer format. The 32-bit integer format uses 31 bits to store the number and one bit to store the sign of the number. As we will see, the integer format is not the default in R. When working with numbers, integers are usually coerced into the floating point numeric format, described later. But the basic binary format underpins the storage of numerical data.[1]

The integer data type cannot store any fractional part of the number. Within these limitations, not many numbers can be represented. The largest positive number R can hold as an integer is $2^{31} - 1$ or 2147483647. As we increment the number by one, we quickly reach a number too large for the data type:

```
> as.integer(2^31 - 2)
[1] 2147483646
> as.integer(2^31 - 1)
[1] 2147483647
> as.integer(2^31)
[1] NA
```

We use the `as.integer` function to ensure the result is converted into an integer. R, if left to its own devices, converts everything to numeric, which we will discuss in Section 2.2.2. At the negative extreme, we can see the same behavior as we decrement toward the minimum integer:

```
> -2147483646L
[1] -2147483646
> -2147483646L - 1L
```

---

[1]See Knuth (1997) for a background on the binary number system, other bases, radix conversion, and other properties of numbers at the hardware-level.

Hexadecimal	Decimal	Binary	Hexadecimal	Decimal	Binary
0	0	0000	8	8	1000
1	1	0001	9	9	1001
2	2	0010	A	10	1010
3	3	0011	B	11	1011
4	4	0100	C	12	1100
5	5	0101	D	13	1101
6	6	0110	E	14	1110
7	7	0111	F	15	1111

Table 2.1
Hexadecimal, decimal, and binary numbers with equivalents

```
[1] -2147483647
> -2147483646L - 2L
[1] NA
```

Here, instead of the `as.integer` function, we can append capital letter "L" to the number to force the integer format. The calculation will continue using integers if all component parts are integers. The letter L stands for *long*, the name of the declaration for a 32-bit integer. Removing the final L converts the intermediate stages to numeric.

```
> -2147483646L
[1] -2147483646
```

Generally, we will call these the maximum and minimum of the data type, the largest positive and negative numbers. Closely related, we will call the largest number the number with the greatest magnitude, regardless of sign. Finally, the smallest number is the term used to describe numbers of lowest magnitude, regardless of sign. These are the numbers closest to 0. In the case of integers, this is 1 and -1.

The maximum value of a 32-bit integer, roughly 2.1 billion, is too small for many common figures. The 2010 population of the world, just under 7 billion people, could not be represented by a 32-bit integer. The value of 13! is more than 6 billion. And there are approximately 4.33×10^{19} potential arrangements for a Rubik's cube (Bandelow 1982, 45). This also ignores the fact the integer data type cannot store numbers smaller than 1, and if we entered 1/2, it would round down to zero.

```
> as.integer(0.5)
[1] 0
> as.integer(1.9)
[1] 1
```

This is not to say there are no advantages to storing numbers as integers. One is that it provides a convenient format for entering binary numbers indirectly. R does not support binary digit input, without the use of a package library. But it does support hexadecimal input. And we can use hexadecimal input to quickly enter binary because there is an exact 4-bits (binary) to 1 digit (hexadecimal) conversion factor. Table 2.1 provides a reference table for the binary-decimal-hexadecimal digits from 0 to 15. Hexadecimal numbers can be entered if they are preceded by the "0x" code to signal the use of hexadecimal numbers:

```
> 0xFACE
[1] 64206
```

Of course, R has no trouble with numbers larger than $2^{31} - 1$. However, it will not use the internal integer storage format to hold these numbers. In fact, the intermediate forms of 2^{31} used in the prior two examples were not integers. Accordingly, these expressions were valid and held the full value until the final conversion to the integer data type.

```
> 2^32
[1] 4294967296
> class(2^32)
[1] "numeric"
```

In the example above, the numbers are being stored in the internal numeric data type, which supports numbers larger than a 32-bit integer can store. Numeric, based on the underlying floating point number format, provides the most flexible and powerful option for native mathematics available to R.

2.2.2 Floating Point Numbers

Floating point numbers provide the way around the limitations of binary integers. Floating point numbers are capable of storing noninteger values, such as 2.71828182845905, 3.14159265358979, and 0.25. Floating point numbers are also capable of storing much larger numbers, as you can see here:

```
> 2^31
[1] 2147483648
> 2^40
[1] 1.099512e+12
```

Floating point numbers also include large magnitude negative numbers. Without the assignment to the integer data type, R defaults to storing the numerical data as floating point data. There are several standards for floating point, but we will focus on the double precision floating point number, usually just called a *double*.

The double corresponds to the C and Java programming language data type, `double`. It is called double precision because it has approximately double the storage space available than the standard floating point number format, corresponding to the `float` data type from C. The double format is the default storage format and is used when R identifies a number format as numeric.

Unless you have specifically changed the specification of a number, as we did above with the `as.integer` function, R is almost certainly using numeric to store data. The double precision data type has 64 bits of data storage per number.

The C `float` data type is sometimes called single precision, and R includes a function to convert a number to a single precision value. However, the underlying data type within R is still double precision and the function has no actual effect. The R function provides internal bookkeeping for R's interactions with languages that support single precision data types.

The double precision data type is based on Institute of Electrical and Electronics Engineers (IEEE) 754, the standard for storing floating point numbers.[2] The IEEE standard provides for a noncontinuous space representing both very large and very small numbers. Under the standard, each floating point number is composed of three parts: the base, exponent, and mantissa. It functions just like scientific notation, but the base is not necessarily 10.

For traditional scientific notation, the base is 10, because humans are used to working with numbers in base 10. For double precision numbers, the base is not encoded in the storage. The base is 2 and is implicit. All IEEE double precision, and therefore, numeric, data types use base 2. Accordingly, it would be redundant to store that information with each instance of a double precision number.

The primary part of the number is called the *mantissa* or *significand*. This number encodes the significant digits of the number represented in the base. Since all double precision numbers are in base 2, this is binary. In this case of a double precision or numeric data type, 53 of the 64 bits are used to store the mantissa, with one of those bits reserved for the sign of the mantissa.

The final part of a floating point number is the exponent. The exponent, like the exponent on a number recorded in scientific notation, is the power the base is raised to, which is then multiplied by the mantissa for a final number. For a double precision data type, there are 11 bits available for storing the exponent, with one bit reserved for the sign of the exponent.

With these three components, a floating point number like 1000 is stored as,

$$1000 = 0b1111101000 * 2^9, \tag{2.4}$$

where the 2 is never stored and $0b\cdots$, shows the binary representation of a number. Similarly, a negative exponent can represent fractional values. Consequently, the number 0.75 is also representable as,

$$0.75 = \frac{3}{4} = 0b0001 * 2^{-1}, \tag{2.5}$$

as a floating point number. There is only one digit included because the initial

---

[2]Severance (1998) provides an interview with William Kahan, one of the standards creators, explaining his motivations and design decisions. See Zuras et al. (2008) for the standard.

digit is implicit. Therefore, this number is 0b0.11 in binary. Because the number is a fractional part in base 2, any number can be represented as the sum of fractions where the denominator is a power of 2. The fraction 3/4 can be represented by $1/2 + 1/4$. As binary decimals, $1/2 = 0b0.1$ and $1/4 = 0b0.01$ This adds up and $3/4 = 0b0.11$. Numbers that are not the sum of such fractions cannot be represented at the exact value in the double precision format and are instead approximated. One common example of this is $1/3$.

Storing numbers in the double precision format has advantages for the numerical processor that handles these numbers. For addition or subtraction, both numbers should be in a common exponent. One can be converted to the other by changing the mantissa. After this conversion, the numbers are added or subtracted, then the result stored as a new floating point number. Using 10 as a base, we can add 200 to 2000 and watch the process:

$$\begin{aligned}200 + 2000 &= 2 \times 10^2 + 2 \times 10^3 \\ &= 2 \times 10^2 + 20 \times 10^2 \\ &= 22 \times 10^2 \\ &= 2.2 \times 10^3\end{aligned}$$

The double precision format is equally adept at handling multiplication and division. For neither is changing the exponent necessary. To multiply two numbers, the mantissae are multiplied together and the exponents are added together. For division, the mantissae are divided and the exponents are subtracted. Again, using 10 as a base for its relative familiarity,

$$\begin{aligned}200 * 2000 &= 2 \times 10^2 * 2 \times 10^3 \\ &= (2 * 2) \times 10^{2+3} \\ &= 4 \times 10^6\end{aligned}$$

Generally, the internal details of how floating point operations work are not necessary for numerical analysts. But some specific data about the double precision floating can help some analyses before they become too big and complex.

In addition to representing numbers, floating point numbers permit several special values to be stored. The most important of these is NaN, represented in R by "NaN." NaN stands for "not a number." One way to get a result of NaN is $0/0$:

```
> 0 / 0
[1] NaN
```

NaN has the property that if introduced into an expression, the NaN will generally propagate throughout the rest of the expression. For instance, $\text{NaN} + x = \text{NaN}$, for any value of x. Also, $\text{NaN}x$, $\text{NaN} * x$, and NaN/x, all equal NaN for any value of x, including if x is itself NaN. There is an exception. Under the IEEE standard, $\text{NaN}^x = \text{NaN}$ unless $x = 0$, in which case $\text{NaN}^0 = 1$. There is a test function, `is.nan` that tests if the value given is NaN. Like the other `is.*` functions, it returns a value of TRUE or FALSE and can accept vector and matrix arguments. Finally, NaN is not equal to anything, even itself:

```
> NaN == NaN
[1] NA
```

NaN is distinct from NA. NaN implies a result that cannot be calculated for whatever reason, or is not a floating point number. Some calculations that lead to NaN, other than 0/0, are attempting to take a square root of a negative number, or perform calculations with infinities that lead to undefined results:

```
> sqrt(-1)
[1] NaN
> Inf - Inf
[1] NaN
```

However, adding two infinities produces ∞:

```
> Inf + Inf
[1] Inf
```

NA is different from NaN in that NA is not a part of the IEEE standard for floating point numbers. NA is a construction of R used to represent a value that is not known, as a placeholder. NA says no result was available or the result is missing. It can be used in a matrix to fill in a value of a vector:

```
> c(1, 2, 3, 4, NA, 5, 6)
[1] 1 2 3 4 NA 5 6
> matrix(c(1, 2, NA, 4, NA, 6, NA, 8, 9), 3)
     [,1] [,2] [,3]
[1,]    1    4   NA
[2,]    2   NA    8
[3,]   NA    6    9
```

In addition, NaN is different from infinity, which has its own representation. Within R, positive infinity is represented as `Inf` and negative infinity is represented by `-Inf`. The value of infinity is the result of a couple of different operations, including exceeding the basic storage requirements. One interesting property of the double precision data type is that division by 0 yields infinity. Within most algebraic frameworks, division by 0 yields an undefined result. We normally ascribe no meaning to this undefined result. It is just undefined, and not infinity, so the definition matches.

Within double precision arithmetic, and by extension, R numeric arithmetic, division by 0 leads to infinity, except when the dividend itself is 0:

```
> 1 / 0
[1] Inf
```

And this is balanced by a certain symmetry that almost any value divided by Inf returns 0:

```
> 1 / Inf
[1] 0
```

Of course, Inf divided by itself is undefined:

```
> Inf / Inf
[1] NaN
```

In addition to Inf, the double precision data type includes a negative infinity, -Inf. This behaves in the same way as Inf, except the sign propagates as expected. Inf also has the property of being equal to itself, but not equal to -Inf. Inf is greater than all other numbers and -Inf is less than all numbers:

```
> Inf == Inf
[1] TRUE
> Inf == -Inf
[1] FALSE
> Inf > 100
[1] TRUE
> Inf < 100
[1] FALSE
> -Inf > 100
[1] FALSE
```

Infinity is also the next larger number than the maximum value of stored double precision number. For instance, if you double the value of .Machine$double.xmax, the number returned is Inf.

Finally, the most surprising aspect of the double precision data type is the existence of a signed zero, or ± 0. Negative zero, -0, acts like 0, but the negative sign tends to propagate as expected:

```
> 1 / -0
[1] -Inf
```

Further, if a result is -0, it will be displayed as 0 while the negative is still carried in the number and used in calculations. This can lead to a somewhat confusing result:

```
> -0
[1] 0
> 1 / -Inf
[1] 0
> 1 / sqrt(-0)
[1] -Inf
```

Otherwise, -0 acts like 0, and $-0 = 0$. Generally, $x - x = 0$ for any value of x, and not -0.

These special values give the floating point system a way to handle exceptional conditions. In some other environments, division by zero, for instance, automatically terminates the program. By instead returning a NaN, floating point allows computation to continue while still making it clear to us that something has gone wrong.

2.3 Numerical Error

In addition to accuracy and precision in our source numbers, the idiosyncrasies of the floating point formats can create new errors. As we have seen, some numbers can be represented precisely while others cannot. This mixing and matching of numbers and storage types can lead to new errors, above and beyond measurement error, that we must be able to plan and account for.

2.3.1 Round-Off Error and Machine ϵ

Another characteristic of number storage we should look at is machine error. The machine error, frequently just called machine ϵ, captures the spacing of numbers within the floating point space. As we have seen, floating point can represent a very broad range of numbers. However, it cannot represent every number between 0 and 1.79769313486232e+308.

Machine ϵ is defined as the smallest value such that,

$$1 + \epsilon > 1, \tag{2.6}$$

within the floating point system. Intuitively, we know that 1 plus any number greater than 0 will be greater than 1. But with the number spacing within floating point, that is not necessarily the case. Machine ϵ is the threshold value for representation.

Given the smallest possible number we found the double precision type to support, we might expect this number to be so minuscule as to be inconsequential. However, this number, available within R as .Machine$double.eps, is comparatively quite large:

```
> .Machine$double.eps
[1] 2.220446e-16
```

While the smallest number is hundreds of orders of magnitude smaller than even the smallest measurable distance, the value of machine ϵ is at the scale of some values we might need. For instance, the width of a proton is 8.4087×10^{-16} meters, a value just about four times larger than machine ϵ. But other subatomic particles can be even smaller and it is possible to inadvertently cross these limits with real-world analyses.

The limitation on machine ϵ is a real limit. Using the print function to force R to print more digits, we can see what happens when we add machine ϵ to 1, along with some nearby values:

```
> print(1 + .Machine$double.eps, digits = 20)
[1] 1.000000000000000222
> print(1 + .Machine$double.eps * 2, digits = 20)
[1] 1.0000000000000004441
> print(1 + .Machine$double.eps / 2, digits = 20)
[1] 1
```

Values equal to or greater than machine ϵ, when added to 1 are distinct. Those smaller are not. The computer will simply round up or down a result appropriately and leave it at that.

R provides a second value to measure the precision of the floating point implementation called .Machine$double.neg.eps. This is the smallest value ϵ such that,

$$1 - \epsilon < 1, \tag{2.7}$$

within the floating point system. Like .Machine$double.eps, we expect any positive number should meet this requirement, but for the same reasons, this is not the case. This number is also much larger than some applications may call for:

```
> .Machine$double.neg.eps
[1] 1.110223e-16
```

And we can see the number in action, just like with .Machine$double.eps.

```
> print(1 - .Machine$double.neg.eps, digits = 20)
[1] 0.99999999999999988898
> print(1 - .Machine$double.neg.eps * 2, digits = 20)
[1] 0.99999999999999977796
> print(1 - .Machine$double.neg.eps / 2, digits = 20)
[1] 1
```

Because floating point is a scientific notation system, the relative effect of different addends to different numbers can change. We will define a new value, ϵ_x, such that,

$$x + \epsilon_x > x, \tag{2.8}$$

in the floating point system where x is any positive number. As x increases in magnitude, the value of ϵ_x also increases. We can observe this experimentally by adding ϵ to, for instance, 1000,

```
> print(1000 + .Machine$double.eps, digits = 20)
[1] 1000
```

which has no effect. The value of ϵ_x is directly related to, and proportional to, the value of x. A similar relationship can be shown for the corresponding negative ϵ for x. The library pracma contains a function, eps, that provides the value of ϵ_x for any given value of x:

```
> library(pracma)
> eps(1000)
[1] 1.136868e-13
> eps(1000000)
[1] 1.164153e-10
> eps(1000000000)
[1] 1.192093e-07
```

As we can see, the value of ϵ_x grows with x and the magnitude of ϵ_x has a linear relationship with the magnitude of x.

The most important effect of this is that certain numbers cannot be represented precisely within a floating point system. It is obvious that transcendental numbers, such as π or $\sqrt{2}$, require infinite precision. Since infinite precision requires infinite storage capacity, we can quickly assume these numbers must be represented by approximations rather than true values.

This is a machine-induced error called *round-off error*. This error is the difference between the true value x and the floating point representation thereof, x'. Round-off error is a manifestation of accuracy problems and can lead to some bizarre interactions.

Transcendental and irrational numbers are not the only unrepresentable numbers. Some perfectly rational numbers such as $1/3$, which repeats infinitely, cannot be represented. And in binary, even $1/10$ repeats infinitely. These numbers are represented by rounded off approximations. Finally, even numbers that do not repeat in binary, and are rational, cannot be represented if the mantissa requires more than 53 bits for storage, including the sign.

Sometimes, this is a problem, but we can counter it with our intuitive understanding of mathematics. For instance, we usually see the number $0.333\ldots$ and convert it to $1/3$ in our heads. But if we see 1.28571428571429, we probably do not immediately recognize it as $9/7$. As numerical analysis is the search for an answer that is good enough for the application domain, we try to balance the need for a true answer with an answer that meets the need.

Functions within the `MASS` library can convert decimals to reduced fractions. If double precision is insufficient to meet those requirements, several packages for R are available that can provide greater precision. The first is `gmp` which provides multiple precision floating point arithmetic. Based on a widely available library, also called `gmp`, this multiple precision library is well tested and provides precision to whatever finite degree we ask. Further, the library `Rmpfr` provides additional function using the `gmp` library as a base.

2.3.2   Loss of Significance

While round-off error happens in or near the last place of numbers, round-off error can appear in other places. Round-off error can be brought to prominence with a simple subtraction exercise:

```
> 1 / 3 - 0.33333333333333
```

[1] 3.330669e-15

Here, $1/3$ is stored as 53 binary digits, which is around 15 decimal digits of significance. Subtracting 0.33333333333333, non-repeating decimal, from $1/3$ should return $3.\overline{3} \times 10^{-15}$. But the floating point representation of $1/3$ has no significant digits after the fifteenth. As a result, the floating point subtraction results in the leftover bits in the representation that are, in reality, the round-off error in the storage of $1/3$.

This problem is called *loss of significance*. We can see the same result subtracting a number near to 1 from 1:

```
> 1 - 0.999999999999
[1] 9.999779e-13
```

Loss of significance will occur when subtracting any two "close" numbers when at least one is not perfectly represented in binary. Mathematically, we note that round-off error becomes more significant in subtraction of the form $x - (x - \delta)$ as δ goes to and for any value of x. This error can intensify if the subtraction is an interim value preceding a multiplication, since the interim result will just be multiplied as is and the error will increase in magnitude:

```
> (1 - 0.999999999999) * 1000
[1] 9.999779e-10
```

In practice, formulae can often be rearranged to remedy these sorts of subtractions, as we saw with Kahan's summation algorithm in Section 1.3.1.

In other cases, round-off error can change intermediate results. For instance, this simple addition problem shows how a simple subtraction problem, $20.55 - 1.35 - 19.2$, is not commutative in floating point arithmetic:

```
> 20.55 - 19.2 - 1.35
[1] 1.332268e-15
> 20.55 - 1.35 - 19.2
[1] 0
```

When we perform both of these calculations by hand, the value is 0. But a simple rearranging to allow for the way mathematics is done by the computer changes the result from something close enough to 0 that we probably do not care, to a value that is truly 0.

Loss of significance is characterized by the behavior that the change in relative error of a result, following an operation, is much larger than the change in absolute error. As explained in Section 2.1.1, we use relative error to understand the effect of the error, by showing its magnitude relative to the final result. Generally, the relative error should be insignificant compared to the absolute error. Loss of significance flips this around by creating a situation where the change in absolute error is very small, comparatively. This hinders our ability to understand the error.

Fortunately, we can solve many instances of loss of significance by changing our approach to the calculation. Using the familiar quadratic formula, we can

```
quadratic <- function(b2, b1, b0) {
    t1 <- sqrt(b1^2 - 4 * b2 * b0)
    t2 <- 2 * b2

    x1 <- - (b1 + t1) / t2
    x2 <- - (b1 - t1) / t2
    return(c(x1, x2))
}
```

R Function 2.1
The quadratic formula

demonstrate this. Given a, b, and c, such that $ax^2 + bx + c = 0$, the quadratic formula is,

$$x = \frac{-b \pm \sqrt{b^2 - 4ac}}{2a}, \tag{2.9}$$

which is implemented in R using the naïve approach in Function 2.1.

The quadratic formula is a classic example of potential loss of significance. The formula itself includes a single subtraction that is, following a square root operation, subject to a division. This combination can yield errors when the subtraction's terms are of the same magnitude, which may be the case with $b^2 - 4ac$.

For some quadratic equations, this is a nonissue. If the equation in question is $24x^2 - 50x - 14$, then

$$b^2 - 4ac = 50^2 - 4(24)(14) \tag{2.10}$$

$$= 2500 - 1344 \tag{2.11}$$

$$= 1156 \tag{2.12}$$

And the square root of 1156 is 34 and nothing terribly interesting has happened. This is usually the case and there are a number of contrived examples of loss of significance that take place in an imaginary floating point calculator, of five or ten significant digits. It is possible to avoid these by changing the algorithm to not perform any subtraction in the numerator. This eliminates the catastrophic cancellation that can cause problems.

For the quadratic formula,

$$x = \frac{-b \pm \sqrt{b^2 - 4ac}}{2a} \tag{2.13}$$

$$= \frac{-b \pm \sqrt{b^2 - 4ac}}{2a} \times \frac{-b - \sqrt{b^2 - 4ac}}{-b - \sqrt{b^2 - 4ac}} \tag{2.14}$$

$$= \frac{2c}{-b \mp \sqrt{b^2 - 4ac}}. \tag{2.15}$$

```
quadratic2 <- function(b2, b1, b0) {
    t1 <- sqrt(b1^2 - 4 * b2 * b0)
    t2 <- 2 * b0

    x1 <- t2 / (-b1 - t1)
    x2 <- t2 / (-b1 + t1)

    ## Reverse the order so they come
    ## back the same as quadratic()
    return(c(x2, x1))
}
```

R Function 2.2
The quadratic formula rewritten

An example of this is provided in Function 2.2, which is mathematically equivalent to the quadratic formula.

In practice, double precision numeric calculations provide sufficient significance, that these results are not common. However, it is possible to experience catastrophic cancellation. In one more extreme case, assume that,[3]

$$a = 94906265.625, \tag{2.16}$$
$$b = 189812534.000, \text{ and} \tag{2.17}$$
$$c = 94906268.375. \tag{2.18}$$

These values, under double precision arithmetic, yield incorrect results, though we must use the **print** function to see the higher precision in R:

```
> b2 <- 94906265.625
> b1 <- 189812534.000
> b0 <- 94906268.375
> print(quadratic(b2, b1, b0), digits = 20)
[1] -1.0000000144879792607 -1.0000000144879792607
```

The correct result is 1 and 1.000000028975958..., though even under double precision arithmetic, R is unable to calculate it correctly using the quadratic formula. Not every proposed fix can solve each problem. For instance, we can use quadratic2 with this example, as well:

```
> print(quadratic2(b2, b1, b0), digits = 20)
[1] -1.0000000144879790387 -1.0000000144879790387
```

---

[3]This example was discovered by Kahan (November 20, 2004), who provides substantially more detail on the subject. In addition, Kahan's work on loss of significance is remarkable, leading to the Kahan summation algorithm from Section 1.3.1.

Here, we have a different incorrect result that is not, in any meaninful way, better than the results we received from the `quadratic` function.

2.3.3   Overflow and Underflow

Because R's numeric data type can hold such a wide range of numbers, underflow and overflow are less likely to be a problem than they could be with other systems. However, it is good to understand the risks of underflow and overflow when R is working with other operating environments.

Underflow is the phenomenon that a number is too small, regardless of sign, to be represented by the computer as anything other than 0. Double precision numeric data types, as shown, can represent numbers so small, that the number is not realistically usable in most applications. Like other machine characteristics, it is available for inquiry, through the `.Machine` object:

```
> .Machine$double.xmin
[1] 2.225074e-308
```

This number, if printed directly, would be a decimal point followed by almost 300 zeros before the first nonzero number appeared. This number is significantly smaller than almost all numbers that may be used in practice. For comparison, the Planck length, the smallest theoretical measurable distance, is $1.61619926 \times 10^{-35}$ meters (Garay 1995). The lowest density vacuum ever created in a laboratory was approximately 1×10^{-18} Pascals (Benvenuti and Chiggiato 1993). And the probability of rolling die 20 times and getting the same number each time is approximately 2.74×10^{-16}.

Even at the extreme lower boundary of small-scale physics and probability, the numbers involved are far larger than the lower bound of double precision. These exceptionally small numbers are, at best, 270 orders of magnitude larger than the smallest number a double precision number can hold. But they are there if we need them down to anything between 0 and `.Machine$double.xmin`.

Even an attempt to contrive an example to illustrate the point is foiled by the floating point standard. The simplest way to actually underflow should be to divide `.Machine$double.xmin` by 2, a result which should yield 0.

```
> .Machine$double.xmin / 2
[1] 1.112537e-308
```

In this case, the machine has outwitted us, again. The internal floating point representation has a method to store even smaller numbers than `.Machine$double.xmin` by moving leading zeros to the mantissa. These numbers are called *denormalized* and provide a method to push the underflow boundary even lower than `.Machine$double.xmin`. However, the implementation is machine dependent and may not necessarily be available. In practice, we are unlikely to reach the underflow boundary with a real-world problem.

Finally, we should keep in mind the `.Machine$double.xmin` value is substantially smaller than machine ϵ. As shown in equation 2.8, the value of the next floating point number after x changes based on how large x is. In fact, the floating point numbers are closer together closer to 0. Further, there are approximately the same number of floating point numbers between 0 and 1 as there are greater than 1.

At the other end of the spectrum, when the number is too large to be expressed, we call the phenomenon *overflow*. Like underflow, overflow is largely a theoretical problem, though one that is easier to create an example for. The largest value can be interrogated in R using .Machine$double.xmax:

```
> .Machine$double.xmax
[1] 1.797693e+308
```

Because the same bits that hold negative exponents also hold positive exponents, this is a number followed by more than 300 zeros. This number is significantly larger than almost all numbers that might be used in practice. For comparison, cosmologists estimate there are only 10^{80} hydrogen atoms in the entire Universe. There are $2^{256} \approx 1.15792089 \times 10^{77}$ encryption keys possible with the 256-bit variant of the AES encryption system. And the surface area of the Milky Way's galactic disk is roughly 10×10^{41} square meters.[4]

Like the smallest numbers, even at the extreme upper boundary of practical mathematics and physics, the numbers involved are far smaller than the upper bound of double precision. These exceptionally large numbers are more than 200 orders of magnitude smaller than the largest possible double precision numbers can store. And like at the smallest numbers, we have larger numbers available, if we need them, up to .Machine$double.xmax.

Here we can better contrive an example to illustrate the point. First, limiting the values to those stored as integers, doubling the value will result in integer overflow:

```
> 2147483647L * 2L
[1] NA
```

And a warning may be displayed, as well.

Of course, without the specification that these numbers be conducted with 32-bit integers, R will dutifully accept and calculate a result:

```
> 2147483647
[1] 2147483647
```

An advantage to using double precision data types to store numbers is that they can store integers up to 2^{53} without any loss of precision. Even though the computer has a mechanism to handle numbers below .Machine$double.xmin, there is no comparable method to store numbers larger than .Machine$double.xmax. Multiplying .Machine$double.xmax by 2 yields a curious result:

---

[4]Both this value and the number of atoms in the Universe are drawn from Wikipedia pages. Neither is properly cited there, either.

```
> .Machine$double.xmax * 2
[1] Inf
```

While we know this number is not infinity, R returns `Inf` because it has no other value available to signify the overflow has occurred.

In both the case of underflow and overflow, R does not cause an error. It simply uses the best available option. In the case of underflow, using 0 may likely be a good enough answer. In the case of overflow, it is unlikely that infinity is a good enough answer, but if seen, should appropriately signal that something has gone wrong.

For the same reasons as underflow, real data is unlikely to cause an overflow condition. However, both overflow and underflow can arise in intermediate calculations. This happens when all the input variables are representable as is the true final answer. However, some intermediate calculation, such as an intermediate sum when calculating a mean, is larger than can be represented.

In addition, not all systems will process data with double precision floating point numbers. R is capable of interacting with and exchanging data with systems that only support single precision arithmetic, where the limits are substantially smaller. In a single precision environment, the smallest possible number is $5.87747175411144e - 39$ and the largest possible number is $3.40282366920938e + 38$. While these numbers are smaller and larger than most applications, we have seen examples where they are insufficient.

We should keep in mind where the boundaries are and be aware if they are approaching. It is possible that we may reach a boundary during an intervening step in a complex calculation. Monitoring and understanding the calculation's complete process will allow us to avoid those conditions, or at least be aware that a result may not be trusted. We will examine that in the next subsection.

2.3.4   Error Propagation and Stability

Each of these error types can manifest at any time. For a single arithmetic operation, this error is probably well below the threshold of concern. But we do not usually use the computer like a pocket calculator. We use the computer to solve more complicated problems that require a lot of mathematics.

Some operations, like matrix transformations, require numerous calculations in succession in order to reach a result. Each of those operations is subject to numerical error and those errors can compound in a form of *cascading error*. Understanding how errors in one calculation affect other calculations in a complex operation is known as *error propagation*.

Using a practical example, rather than one from machine arithmetic, puts this within grasp. Imagine a traditional ruler with marks every millimeter.[5] There is a natural limitation of precision where no measurement of greater precision than a millimeter is possible. A length measurement, ℓ will necessar-

---

[5]You may, if necessary, use sixteenths of an inch as a substitute for the millimeter here; the mathematics works out identically.

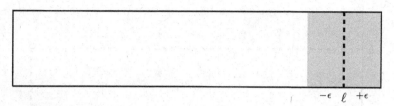

Figure 2.4
Error in the measurement of length

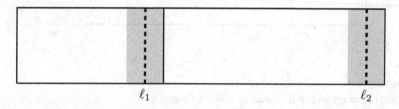

Figure 2.5
Additive error in measurement of length

ily include some error. We will say that $\ell = \ell' \pm \epsilon$ where ℓ is the true length, ℓ' is the measured length, and ϵ is some error. In this case, ϵ is 0.5 millimeters. An example is shown in Figure 2.4 where the potential error area is shaded.

We can allow that this measured error is acceptable because if it were not, we would use a better measuring tool. However, even with that allowance, it can cause trouble. If given two measurements and we must add them, the errors themselves are additive. Under the same circumstances, we have two measurements, ℓ_1 and ℓ_2 such that,

$$\ell_1 = \ell_1' \pm \epsilon \tag{2.19}$$

$$\ell_2 = \ell_2' \pm \epsilon. \tag{2.20}$$

Adding the values together leads to a combined error of,

$$\ell_1 + \ell_2 = (\ell_1' \pm \epsilon) + (\ell_2' \pm \epsilon) \tag{2.21}$$

$$= \ell_1' + \ell_2' \pm (2\epsilon). \tag{2.22}$$

Assuming each measurement were taken the same way, $\epsilon_1\epsilon_2$, but the error grows, regardless. In the example with the ruler, the combined error of two measurements added is one millimeter. This is shown in Figure 2.5 where the potential area, subject to the error estimates, is shaded. Accordingly, they grow linearly and if we were designing a container to hold ten objects of length ℓ, the combined error is 10ϵ, and so errors can propagate through an addition process.

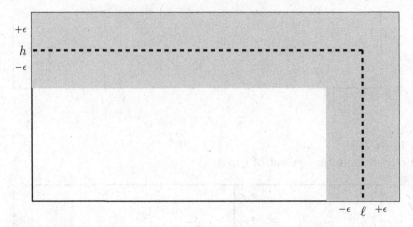

Figure 2.6
Multiplicative error in measurement of length

Other situations can amplify error more extremely. Imagine two measurements, a length, ℓ and a height, h, with a desire to find the surface area. If both were taken via the same method and have the same potential error in the measurement, ϵ, the error is multiplicative, across all terms, and,

$$\ell \times h = (\ell' \pm \epsilon) \times (h' \pm \epsilon) \tag{2.23}$$
$$= \ell' h' \pm (\ell' \epsilon + h' \epsilon + \epsilon^2). \tag{2.24}$$

The entire error in the surface area is $\ell' \epsilon + h' \epsilon \epsilon^2$. If the measurement error were assumed to be within half a millimeter, as the prior example, then the combined potential error in surface area is at more than half the length and height added together, in square millimeters. This is shown graphically in Figure 2.6. The potential error in a volume measurement is, commensurately, cubical.

Through these examples, we can see the affects of compounding errors. When a value is subject to machine error, through the floating point representation or through some other source, the same error propagation patterns emerge. In the case of the surface area shown above, the calculation will yield a result, even if it is not necessarily correct.

But if one of the source values, such as the height, had overflowed, and was now represented as `Inf`, then the final result would be `Inf`. The same thing can happen with `NaN`. While these results are obviously unnatural, they arise just as easily through mistake as they do through actual results. An accidental logarithm of 0 returns `-Inf` and if buried within a larger calculation, we can be left scratching our heads for a few minutes as we look for what went wrong.

This leads to the search for numerical stability, another important goal of numerical analysis. *Numerical stability* is the property of numerical algorithm that the approximation errors, from whatever source, are not amplified by the algorithm. This property manifests in two important ways.

The first way it manifests is that numerically stable algorithms will not produce drastic output changes after small changes in the input. Drastic changes are sometimes in the eye of the beholder. An algorithm where the output is an exponential function of the input will grow quickly as the input value increases. Here, a drastic change is one beyond the expectation of change.

The second manifestation of numerical stability refers to how the algorithm behaves near certain boundary values or singularity points. An algorithm that divides by the input will behave well until the input value is 0, division's singularity. However, how the algorithm behaves near the singularity is expected and only produces incorrect results at 0.

A numerically unstable program is not well behaved in this sense. One example would be an algorithm that will magnify the inherent errors without bound. An analysis of the program will show that the final error is indeterminate, or if determined, the relative error is large, leading to unusable results. Some algorithms are only unstable at certain boundary points, when a component function has no return value or meaning. In these cases, the algorithm may be stable for some input domains, but not others.

Numerical stability is a desirable property to have in an algorithm, but it is generally found in an algorithm-problem pairing. As a result, it does not necessarily follow that an algorithm-problem pairing which is stable will remain stable if the problem is changed. Sometimes, problem areas can be identified ahead of time. In other cases, instabilities are best observed experimentally. Observing how an algorithm behaves and produces results for different inputs is a key aspect of numerical analysis.

2.4 Applications

With a firm understanding of how data is stored in R, we will now explore the first application of this knowledge. In Section 2.4.1, we will see a simple method to perform integer division, with a remainder, through repeated subtraction. This method is slow and not very interesting. Now, using the mechanics of binary data, we can explore a new method that implements long division like we learned in school.

2.4.1 Simple Division Algorithms

In our exploration of fundamental problems and efficiency, we begin by exploring two algorithms which should not be used in practice. These algorithms perform basic division on integers returning the quotient and a remainder. This is usually called Euclidean division. These algorithms should not be used

```
naivediv <- function(m, n) {
    quot <- 0
    r <- m

    if(n == 0)
        stop("Attempted division by 0")

    while(r >= n) {
        quot <- quot + 1
        r <- r - n
    }

    return(list(quotient = quot, remainder = r))
}
```

R Function 2.3
Naïve division by repeated subtraction

as a replacement for the internal R operations for division and modulo, integer division that just returns the remainder. However, they do perform the operations and can be implemented in just a few lines of R.

The simplest approach to division mimics the basic method of multiplication. When learning multiplication, we learn the times tables. But before that, we learn to intuit the mechanism by successively adding blocks of numbers together. Given four sets of three apples each, we learn to multiply by adding four sets of three together, $4 \times 3 = 3 + 3 + 3 + 3 = 12$. We can divide by reversing the process.

This method is provided in Function 2.3, the function `naivediv`. We will describe this function, and others, as naïve, but this should not be considered a pejorative. The naïve approach is the most basic approach you could take to solving a problem. The `naivediv` function accepts two options. The first is the dividend, or the numerator of the division process. The second is the divisor, or the denominator.

The function begins by setting an internal value, r to the dividend. From there, the function subtracts the denominator from r in a loop until the value of r is less than the divisor. For each subtraction, the value of a second internal value, q, is incremented. When r is less than the divisor, the value of *quot* is the quotient and r contains the remainder. The function returns a list with two components named *quot* and r representing each value, respectively.

The algorithm is, at best, inefficient. It will cycle through the loop $q = \lfloor m/n \rfloor$ times. That is quite a bit and while it reduces when n is larger, it increases when m is larger. The function returns the quotient and remainder for any positive values of m and n:

```
> naivediv(314, 7)
```

```
$quotient
[1] 44

$remainder
[1] 6
```

Before using these division functions, remember that R has a built-in operator that performs the modulo operation, `%%`. The `floor` function, combined with basic division, should be used to get the quotient. The modulo operator should be the preferred choice:

```
> floor(314 / 7)
[1] 44
> 314 %% 7
[1] 6
```

Like most algorithms in this text, there are cleaner, faster, and more robust implementations already available within R.

2.4.2   Binary Long Division

When we are learning division, we only use the inverse multiplication method early and for small numbers. We also use the standard multiplication tables as a mental shortcut. For larger numbers, we use the long division, dreaded by elementary students for generations. The long division algorithm works by breaking the dividend into its components, at the level of its digits, and dividing each set of digits. Revisiting the last example, the long division for 314/7 is:[6]

$$
\begin{array}{c|c}
3\,1\,4 & 7 \\ \cline{2-2}
-\ 2\,8 & 4\,4 \\ \cline{1-1}
\ \ 3\,4 & \\
-\ \ \ 2\,8 & \\ \cline{1-1}
\ \ \ \ \ 6 &
\end{array}
$$

Of course, as the dividend is broken up into the number of digits as are in the divisor, with each step:

$$
\begin{array}{c|c}
3\,2\,7\,6\,7 & 1\,0\,2\,3 \\ \cline{2-2}
3\,0\,6\,9 & 3\,2 \\ \cline{1-1}
\ \ 2\,0\,7\,7 & \\
-\ 2\,0\,4\,6 & \\ \cline{1-1}
\ \ \ \ \ 3\,1 &
\end{array}
$$

The algorithm is typically conducted in base 10, but we can replicate this process in any base. The function `longdiv`, provided in Function 2.4,

---

[6]With apologies to the American readers, but as differing tableau styles for long division are available via LATEX, the French style is much clearer to follow.

```
longdiv <- function(m, n) {
    quot <- 0
    r <- 0

    if(n == 0)
        stop("Attempted division by 0")

    for(i in 31:0) {
        r <- bitwShiftL(r, 1)
        r <- r + bitwAnd(bitwShiftR(m, i), 1)
        if(r >= n) {
            r <- r - n
            quot <- quot + bitwShiftL(1, i)
        }
    }

    return(list(quotient = quot, remainder = r))
}
```

R Function 2.4
Binary long division

does this. This function, after preparing an initialization for the quotient and remainder, executes a simple loop over the bits of the dividend. The function need only execute over the significant bits and eliminate leading 0s; however, when executing the algorithm, it simply returns 0 for each of those bits.

The algorithm uses the bit manipulation functions inside of R to provide access to the bits. First, shifting the bits one unit to the left is equivalent to multiplying by two. A shift to the right is equivalent to dividing by two, with the remainder lost. The quotient is built up bit by bit, with the highest order bit first; any remainder, at each point, is carried through to the next step.

```
> longdiv(314, 7)
$quotient
[1] 44

$remainder
[1] 6
```

The longdiv is substantially faster than division by repeated subtraction. Further, its runtime is bounded by the size of the dividend, not both the dividend and the divisor. In this implementation, the loop is executed over all the available bits of storage, regardless of whether they are used. Therefore, the loop will execute exactly 32 times, sufficient for the 32 bits of any integer in R. For 1000/3, the naivediv function executes the loop 334 times, ten times the maximum of longdiv.

Comments

Understanding error, its nature and sources, is a key part of numerical analysis. There are two main types of error we must contend with, round-off and algorithmic, and they manifest in different ways and for different reasons. Worse, they can appear individually or in concert with each other. However, each type of error can be mitigated through defensive programming techniques and the knowledge that, no matter how hard we try, error cannot be eliminated. It can, at best, be bounded, in the general case.

Further, the use of double precision arithmetic allows us to save some cases of round-off error before they happen. Round-off error, in a double precision number, is sufficient to solve most numbers we need and solve most problems we have. And there is good news in this.

Most numerical analysis problems represent some form of applied mathematical problem. Perhaps it is defining a curve for a wind turbine or estimating wealth disparity. Applied mathematical problems generally involve some type of measurement of real-world values. In practice, our measurement error will be a greater limiting factor for many of the calculations we will perform. For instance, a standard rule will only have a measurement resolution of 1/16th of an inch, or so. That number can be stored as a floating point number without problems. However, with real-world data, each of the individual measurement errors is a range that is added or subtracted to the given measurement. If we are lucky, these errors cancel each other out leaving us with a net-zero error. In practice, when measurement error is unbiased, then the error is centered at 0 and appears with an estimable frequency pattern.

But a measuring tool, such as a ruler, may be biased, one way or the other. Consequently, especially in the realm of error propagation, the defined errors are maximum potential error, not minimum and certainly not estimates of the actual error. It is obvious to think that if we know what the error is, we can just adjust our results to accommodate it. But in reality, all we have is an upper bound, in most cases.

As we work with numerical analysis, we want to keep an eye on that upper bound of the error. The upper bound is still useful. We can use that to manage expectations on results. Many of the algorithms we work with are focused on minimizing that upper bound. When we fit that upper bound within a tolerance we have specified, we know that our answer was good enough, for whatever good enough means at that time. Fitting our algorithm to meet our needs is a crucial part of the selection process, on par with meeting the requirements we specified in Chapter 1.

The rest of this book will now introduce a series of algorithms for solving different types of mathematical problems. Some are general solutions and some are keyed to specific problems. Finally, some will be applied to specific real-world problems as we see how all of this connects back to applied mathematics.

Exercises

1. What is the binary representation, as an integer, of the number 4714?

2. What is the floating point representation of the number 8673.34?

3. What is the floating point representation of the number 1/6?

4. The number e, the base of the natural logarithm, is approximately 2.71. What is the absolute error in that estimate? What is the relative error?

5. Give an example of a measurement which is accurate, but not precise.

6. Give an example of a measurement which is precise, but not accurate.

7. How many numbers are there between 1 and `.Machine$double.xmax` in double precision arithmetic?

8. How many numbers are there between 0 and 1 in double precision arithmetic?

9. If a 5×6 array contains double precision numeric, find the minimum number of bytes necessary to store the array.

10. Write the data storage format for the double precision represented by the numbers 1/2, 1/3, and 1/10.

11. Find the smallest possible ϵ such that $1000 + \epsilon > 1000$.

12. Rewrite $f(x) = \sqrt{x^2 + 1} - 1$ so there is no loss of significance.

13. What the relationship between ϵ_x and ϵ_{-x}?

14. Some computers support quadruple precision arithmetic, using 128 bits to store a number. The exponent uses 15 bits, the fractional part uses 112 bits, and the sign is stored in a single bit. What is the largest and smallest number this can represent?

15. Research other methods of storing numeric data such as binary-coded decimal, fixed-point, or Java's `BigDecimal` and discuss the pros and cons of these methods versus floating point numeric data types.

3

Linear Algebra

3.1 Vectors and Matrices

In Chapter 1, we introduced vectors and matrices as R uses them. This introduction focused on the vector and matrix as data types used by R to store data. If you have worked with R for data science or statistics before, you may have already seen these data types. Perhaps you have used vectors to store time series data. You may have used a matrix to store statistical observations. If your work with R has centered on its data processing and statistics capabilities, this makes sense.

But for a mathematician, the powers of vectors and matrices lie in linear algebra. A two-element vector might represent a complex number or ordered pair. A three-element vector may represent a position in three-dimensional space and a four-element vector could be a quaternion. Matrices usually represent systems of equations. Mathematically, there are many applications for each, ranging from graphics processing to physics to statistical analysis.

3.1.1 Vector and Matrix Operations

To begin, we will assume there are two vectors of length n, $\mathbf{u}$ and $\mathbf{v}$, such that,

$$\mathbf{u} = \begin{bmatrix} u_1 \\ u_2 \\ \vdots \\ u_n \end{bmatrix} \text{ and } \mathbf{v} = \begin{bmatrix} v_1 \\ v_2 \\ \vdots \\ v_n \end{bmatrix}. \tag{3.1}$$

If we add a scalar, a single number, to a vector, then the number is added to each element so that,

$$\mathbf{u} + x = \begin{bmatrix} u_1 \\ u_2 \\ \vdots \\ u_n \end{bmatrix} + x = \begin{bmatrix} u_1 + x \\ u_2 + x \\ \vdots \\ u_n + x \end{bmatrix}. \tag{3.2}$$

And if we add a vector to a another vector, we create a new vector that is the element-wise sum of the two vectors. That is,

$$\mathbf{u} + \mathbf{v} = \begin{bmatrix} u_1 \\ u_2 \\ \vdots \\ u_n \end{bmatrix} + \begin{bmatrix} v_1 \\ v_2 \\ \vdots \\ v_n \end{bmatrix} = \begin{bmatrix} u_1 + v_1 \\ u_2 + v_2 \\ \vdots \\ u_n + v_n \end{bmatrix}. \tag{3.3}$$

And in R, this works in the natural way. First, we will create two three-element vectors. Then, we will add a scalar and add the vectors together.

```
> u <- c(1, 2, 3); v <- c(8, 4, 2); x <- 7
> u + x
[1]  8  9 10
> u + v
[1] 9 6 5
```

Of course, successfully adding two vectors is predicated on their length matching. But if the vectors are not of the same length, R will try to help you by "recycling" the shorter vector. When R recycles the vector, after reaching the end, it will start with it again from the beginning and repeat this process until the longer vector has been processed.

```
> u + c(1, 9)
[1]  2 11  4
```

R will normally provide a warning message that is not replicated here when one vector's length is not an integer multiple of the other's. Recycling can be advantageous and provide a shortcut for handling complex translations. In one sense, when adding a scalar to a vector, the scalar is a one-element vector that is recycled over the longer vector. However, relying on R to produce a warning when adding two vectors of different length will not work to catch problems.

Given how addition works, it is reasonable to assume that other key operators, subtraction, multiplication, and division, work the same way, and they do. This is different from, in particular, MATLAB®, which requires a special notation for element-wise operations on vectors. R takes this to the logical extreme and allows the exponent operator (^) to be used in an element-wise fashion. Like the four basic operators, it will recycle one vector over another and produce a warning if their lengths are not a multiple of each other.

Matrices are similar, but encounter somewhat different semantics than the vector because of their multiple dimensions. We will assume there are two matrices, of size m by n, such that,

$$A = \begin{bmatrix} a_{1,1} & a_{1,2} & \cdots & a_{1,n} \\ a_{2,1} & a_{2,2} & \cdots & a_{2,n} \\ \vdots & \vdots & \ddots & \vdots \\ a_{m,1} & a_{m,2} & \cdots & a_{m,n} \end{bmatrix} \text{ and } B = \begin{bmatrix} b_{1,1} & b_{1,2} & \cdots & b_{1,n} \\ b_{2,1} & b_{2,2} & \cdots & b_{2,n} \\ \vdots & \vdots & \ddots & \vdots \\ b_{m,1} & b_{m,2} & \cdots & b_{m,n} \end{bmatrix}.$$

In R, adding A and B adds the two matrices, element-wise, if and only if the matrices are the same size, so that,

$$A + B = \begin{bmatrix} a_{1,1} + b_{1,1} & a_{1,2} + b_{1,2} & \cdots & a_{1,n} + b_{1,n} \\ a_{2,1} + b_{2,1} & a_{2,2} + b_{2,2} & \cdots & a_{2,n} + b_{2,n} \\ \vdots & \vdots & \ddots & \vdots \\ a_{m,1} + b_{m,1} & a_{m,1} + b_{m,2} & \cdots & a_{m,n} + b_{m,n} \end{bmatrix}. \tag{3.4}$$

R creates a convenience out of vector recycling when adding two vectors. However, there is no obvious extension of vector recycling into two-dimensional matrices. As a result, R does not allow for adding two matrices of different sizes, unless one of them is a vector.

Take the example of $A + x$ where x is a scalar value. For this case, like the case of adding a vector and a scalar, the value of x is added to each element of A. In the case of $A + \mathbf{v}$, the vector $\mathbf{v}$ is added to each *column* of A, element wise, as though the column were a freestanding vector. That means the rules for recycling apply. We can see this in the following examples, using $\mathbf{u}$ from before:

```
> A <- matrix(1:9, 3)
> A + 1
     [,1] [,2] [,3]
[1,]    2    5    8
[2,]    3    6    9
[3,]    4    7   10
> A + c(1, 2, 3)
     [,1] [,2] [,3]
[1,]    2    5    8
[2,]    4    7   10
[3,]    6    9   12
```

However, this can result in an unfortunate misunderstanding with adding vectors of different length. The operation will recycle over the first column, going down. Then, if the vector has any left over elements, it will pick up where it left off at the top of the second column, beginning the process all over again. We can see this clearly in the following examples, by adding a 4 by 4 matrix to a three-element vector and then subtracting out the original matrix, leaving a matrix of the recycled vector:

```
> A + 1
```

```
      [,1] [,2] [,3]
[1,]    2    5    8
[2,]    3    6    9
[3,]    4    7   10
> A + c(1, 2) - A
      [,1] [,2] [,3]
[1,]    1    2    1
[2,]    2    1    2
[3,]    1    2    1
> A + c(1, 2, 3) - A
      [,1] [,2] [,3]
[1,]    1    1    1
[2,]    2    2    2
[3,]    3    3    3
```

These sorts of complicated interactions are best avoided if possible.

Adding two matrices of different sizes results in an error. Assuming we use the same matrix for A as before, we will now add a new matrix, B that is 3 rows by 2 columns. We use the `try` function to capture the error and display it here:[1]

```
> B <- matrix(1:6, 3)
> status <- try(A + B)
> print(status[1])
[1] "Error in A + B : non-conformable arrays\n"
```

Of course, multiplication in particular has a special meaning to both vectors and matrices. R provides an operator called `%*%` that provides matrix multiplication for standard matrices. Remembering that for matrix multiplication to succeed, the number of columns of the left matrix must match the number of rows of the right matrix, the same operator provides the matrix multiplication function. Using the matrix A and B we defined before,

```
> A %*% B
      [,1] [,2]
[1,]    30   66
[2,]    36   81
[3,]    42   96
```

results in a new matrix that is the result of the multiplication, AB. Matrix multiplication is not necessarily commutative, therefore, BA may not necessarily be defined. R responds with an error similar to the addition of different sized matrices seen previously.

For vectors, the effect of matrix multiplication is known as the dot product. This is mathematically equivalent to the row-by-column multiplication

---

[1]This book is automatically generated from R and LATEX code. When executing this statement without the `try` function, building the PDF fails. The first element of the return of `try` is the error's text.

product for matrices when the inner dimension of both matrices is one unit. We call this the *dot product* or the *inner product*. The dot product on vectors can only be executed on two vectors of the same length. Given the definitions of **u** and **v** given in equation 3.1 above, then

$$\mathbf{u} \cdot \mathbf{v} = \sum_{i=1}^{n} u_i * v_i. \qquad (3.5)$$

However, the dot product's result is a scalar, a single number. The result of this operator is a matrix of one row and one column. This is because the operator is actually producing a one-by-one matrix with the result of the matrix multiplication as the single entity. In some regards, this may be indistinguishable, depending on the use case. This operator, syntactically acts just like other operators, like the plus sign. It does not act like a function. You can see an example below using the vectors **u** and **v** defined earlier:

```
> u %*% v
     [,1]
[1,]   22
```

Having worked through a range of vector and matrix products, we might want to do some different operations on our matrices. The function `diag` extracts the elements of the principal diagonal of the matrix given:

```
> diag(A)
[1] 1 5 9
> diag(B)
[1] 1 5
```

Interestingly, the `diag` can also be used to construct diagonal matrices. If the first option is a matrix, the principal diagonal is extracted. If the first option is a vector, then a matrix is constructed. If the second option is a count, the matrix is assembled to that size specification, and with a third option, the rows and columns are separately counted. If the vector length given is not equal to row and column count, vector recycling will be used to flesh out a sufficient number of entries in the matrix:

```
> diag(u)
     [,1] [,2] [,3]
[1,]    1    0    0
[2,]    0    2    0
[3,]    0    0    3
> diag(1, 4)
     [,1] [,2] [,3] [,4]
[1,]    1    0    0    0
[2,]    0    1    0    0
[3,]    0    0    1    0
[4,]    0    0    0    1
```

The `diag` function is an excellent way we can create a matrix with ones on the diagonal.

Many of these functions and operations rely on the matrices being specific sizes, relative to each other. Accordingly, we would like to have a set of tools for interrogating vectors and matrices about their size and other properties. For a matrix, we might like to know the number of rows, and that can be found from the `nrow` functions. We might also like to know the number of columns, which can be found via the `ncol`.

```
> nrow(B)
[1] 3
> ncol(B)
[1] 2
```

However, if either of these functions are applied to a vector, the function will return NULL.

```
> nrow(u)
NULL
> ncol(u)
NULL
```

Generally, for vectors, if we know it is a one-dimensional vector ahead of time, we use the `length` function to determine the number of entries, as we have seen before. If the `length` function were applied to a matrix, the result is the total number of entries in the matrix, or $m \times n$.

```
> length(u)
[1] 3
> length(B)
[1] 6
```

Finally, you can use the `dim` function to return all dimensions of the object as a vector. However, if applied to a vector, the result is NULL.

```
> dim(B)
[1] 3 2
```

R has a few other intrinsic functions for manipulating matrices. In addition, there is a standard package, `Matrix`, that provides functions for working with specialized matrices.

3.1.2 Elementary Row Operations

Matrix operations go beyond cross products and adding vectors. Most of the interesting applications of matrices require the three elementary row operations. The first of these is row scaling. With row scaling, all of the elements of a given row are multiplied by some nonzero value. To implement this in R, we can use row addressing to extract a row, multiply by the constant, then set the new row back into its place in the matrix. This is the simplest of the

```
scalerow <- function(m, row, k) {
    m[row,] <- m[row,] * k
    return(m)
}
```

R Function 3.1
The row scaling algorithm

```
swaprows <- function(m, row1, row2) {
    row.tmp <- m[row1,]
    m[row1,] <- m[row2,]
    m[row2,] <- row.tmp

    return(m)
}
```

R Function 3.2
The row swapping algorithm

row operations, and it requires three steps. And though those steps are simplest, one of the advantages to a functional programming language like R is the ability to name functions. Creating a function called `scalerow` allows the reader to understand what the operation does without having to determine the purpose of an actual row operation, which can grow tiresome quickly.

The implementation of `scalerow` is provided in Function 3.1. In this case, the scaling happens in place from the function's perspective. But R is ensuring that all three independent steps outlined above take place. The three options provided are the source matrix, m, the row number, row, and the scaling factor, k. The revised matrix is returned.

```
> A <- matrix(1:15, 5)
> scalerow(A, 2, 10)
     [,1] [,2] [,3]
[1,]    1    6   11
[2,]   20   70  120
[3,]    3    8   13
[4,]    4    9   14
[5,]    5   10   15
```

Note that at the end of this process, the value of A is unchanged and while the return from `scalerow` is displayed, it is not saved and is lost.

The second elementary row operation is the row swap. Like `scalerow`, the row swap function is given a name for convenient reference more than any specific technical reason. The function is called `swaprows` and its implementation is provided in Function 3.2. The implementation of `swaprows` takes three

```
replacerow <- function(m, row1, row2, k) {
    m[row2,] <- m[row2,] + m[row1,] * k
    return(m)
}
```

R Function 3.3
The row replacement algorithm

arguments. Again, the first argument is the source matrix. The second and third arguments are the index numbers of the rows to be swapped. There is no need to order the index numbers in any particular way. The function itself works by creating a temporary variable that holds the first target's destination row. After the first target is overwritten, the second target is overwritten by the contents of the temporary storage vector.

```
> swaprows(A, 1, 4)
     [,1] [,2] [,3]
[1,]    4    9   14
[2,]    2    7   12
[3,]    3    8   13
[4,]    1    6   11
[5,]    5   10   15
```

The last of the elementary row operations is the row replacement algorithm. The process of row replacement might be misnamed in some sense. The row is not truly replaced, but rather has a characteristic transformation applied to it. With row replacement, a row is replaced with the sum of that row, and another row, whose entries have been multiplied by some scalar value. That is, a source row is multiplied by some constant and the result is added to the destination row. We do not really replace anything. The implementation of the row replacement operation, the **replacerow** function, like the other elementary row operations, exists as a convenient nomenclature for other uses. The function, in a single line, performs the scaling, addition, and assignment, then returns the revised matrix. The function itself takes four arguments. The first is the matrix, the second is the source row index, and the third is the destination row index. The final argument is the scaling factor to apply to the source row. The implementation source is in Function 3.3.

```
> replacerow(A, 1, 3, -3)
     [,1] [,2] [,3]
[1,]    1    6   11
[2,]    2    7   12
[3,]    0  -10  -20
[4,]    4    9   14
[5,]    5   10   15
```

As you can see from the example, the `replacerow` function can be very useful in Gauss–Jordan elimination, as we have used it to successfully zero-out the first column of the third row in example matrix A.

3.2 Gaussian Elimination

In the last section, we explored a suite of tools for manipulating and transforming matrices. Some of these are built in to R and others we developed ourselves. We can use these tools now to solve matrix equations.

3.2.1 Row Echelon Form

The first step toward solving mathematical equations is reducing a matrix to *row echelon form* using *Gaussian elimination*. A matrix is in row echelon form if the matrix meets two conditions. First, rows with only values of zero must be below rows with any nonzero values. Second, the first nonzero entry of any other row must be to the right of the row above it. This gives matrices that look something like,

$$A = \begin{bmatrix} d_1 & a_{1,2} & \cdots & a_{1,n} \\ 0 & d_2 & \cdots & a_{2,n} \\ \vdots & \vdots & \ddots & \vdots \\ 0 & 0 & \cdots & d_m \end{bmatrix} \text{ or } B = \begin{bmatrix} d_1 & b_{1,2} & b_{1,3} & \cdots & b_{1,n} \\ 0 & 0 & d_2 & \cdots & b_{2,n} \\ 0 & 0 & 0 & \cdots & b_{3,n} \\ \vdots & \vdots & \vdots & \ddots & \vdots \\ 0 & 0 & 0 & \cdots & b_{m,n} \end{bmatrix}.$$

In short, the matrix will have an increasing number of zeroed entries in the lower rows. Further, the row echelon form of a given matrix is row equivalent to the given matrix.

The algorithm for Gaussian elimination captures the approach when performing Gaussian elimination by hand. First, for the first row, R_1, if the first column is nonzero, then for each following row, a multiple of R_1 is added to the subsequent row, so as to set the first column of that row to zero. We will start with a sample matrix,

$$A = \begin{bmatrix} 5 & 8 & 6 \\ 5 & 2 & 5 \\ 5 & 2 & 4 \end{bmatrix}. \tag{3.6}$$

With matrix A, the leading entry of each column is 5, which is convenient. For the second row, we add $-1 \times A_1$ to both A_2 and A_3, where A_n is the nth row of the matrix A. Following the application to rows 2 and 3,

$$A = \begin{bmatrix} 5 & 8 & 6 \\ 5 & 2 & 5 \\ 5 & 2 & 4 \end{bmatrix} \sim \begin{bmatrix} 5 & 8 & 6 \\ 0 & -6 & -1 \\ 0 & -6 & -2 \end{bmatrix}. \tag{3.7}$$

```
refmatrix <- function(m) {
    count.rows <- nrow(m)
    count.cols <- ncol(m)
    piv <- 1

    for(row.curr in 1:count.rows) {
        if(piv <= count.cols) {
            i <- row.curr
            while(m[i, piv] == 0 && i < count.rows) {
                i <- i + 1
                if(i > count.rows) {
                    i <- row.curr
                    piv <- piv + 1
                    if(piv > count.cols)
                        return(m)
                }
            }
            if(i != row.curr)
                m <- swaprows(m, i, row.curr)
            for(j in row.curr:count.rows)
                if(j != row.curr) {
                    k <- m[j, piv] / m[row.curr, piv]
                    m <- replacerow(m, row.curr, j, -k)
                }
            piv <- piv + 1
        }
    }
    return(m)
}
```

R Function 3.4
The row echelon matrix form

This process will hang up if the leading entry of the first row is zero. Then, the row would be swapped with an appropriate row, one without a leading zero, in the first column. The leading entry is typically called the pivot column. Then the reduction process would proceed as described above. The process repeats with the first nonzero column of the next row, and so on until there are no rows left.

The implementation of the algorithm, provided in Function 3.4, implements the algorithm slightly differently. First, the algorithm determines if the current pivot column is nonzero. If not, it searches for the next row that includes a nonzero entry in the pivot column. Once found, the initial while loop exists and drops out to an if statement.

We do not have to use this if statement. If i and *row.curr* are equal, then the initial pivot column is nonzero. If they are not, then the initial pivot column was zero and i is the index of a suitable replacement row. Then the rows are swapped. If we do not use the `if`, then the row will be swapped with itself, which is effectively a null operation.

After the decision to swap the rows is made, and the swap executed if necessary, then the following loop uses row replacement to eliminate leading entries in the pivot column for all subsequent rows.

The final result is a matrix in upper triangular form. Using the example of A above,

```
> (A <- matrix(c(5, 5, 5, 8, 2, 2, 6, 5, 4), 3))
     [,1] [,2] [,3]
[1,]    5    8    6
[2,]    5    2    5
[3,]    5    2    4
> refmatrix(A)
     [,1] [,2] [,3]
[1,]    5    8    6
[2,]    0   -6   -1
[3,]    0    0   -1
```

And using an example of a 3-by-4 matrix, we can see the function still works,

```
> (A <- matrix(c(2, 4, 2, 4, 9, 4, 3, 6, 7, 7, 3, 9), 3))
     [,1] [,2] [,3] [,4]
[1,]    2    4    3    7
[2,]    4    9    6    3
[3,]    2    4    7    9
> refmatrix(A)
     [,1] [,2] [,3] [,4]
[1,]    2    4    3    7
[2,]    0    1    0  -11
[3,]    0    0    4    2
```

Matrix row operations, however, apply many repeated operations to the same set of numbers. Individual entries in a matrix may be transformed through row replacement for every row in the matrix. As observed in Section 2.3.4, subtracting very close numbers can create unexpected large errors and Gaussian elimination relies on subtraction with row replacement to complete the elimination process. This can happen even with superficially trivial integer matrices.

```
> (A <- matrix(c(2, 8, 5, 5, 1, 2, 3, 8, 4), 3))
     [,1] [,2] [,3]
[1,]    2    5    3
[2,]    8    1    8
[3,]    5    2    4
> refmatrix(A)
```

```
rrefmatrix <- function(m) {
    count.rows <- nrow(m)
    count.cols <- ncol(m)
    piv <- 1

    for(row.curr in 1:count.rows) {
        if(piv <= count.cols) {
            i <- row.curr
            while(m[i, piv] == 0 && i < count.rows) {
                i <- i + 1
                if(i > count.rows) {
                    i <- row.curr
                    piv <- piv + 1
                    if(piv > count.cols)
                        return(m)
                }
            }
            if(i != row.curr)
                m <- swaprows(m, i, row.curr)
            piv.val <- m[row.curr, piv]
            m <- scalerow(m, row.curr, 1 / piv.val)
            for(j in 1:count.rows) {
                if(j != row.curr) {
                    k <- m[j, piv] / m[row.curr, piv]
                    m <- replacerow(m, row.curr, j, -k)
                }
            }
            piv <- piv + 1
        }
    }
    return(m)
}
```

R Function 3.5
The reduced row echelon matrix form

```
       [,1]          [,2]        [,3]
[1,]     2   5.000000e+00   3.000000
[2,]     0  -1.900000e+01  -4.000000
[3,]     0   1.776357e-15  -1.289474
```

The refmatrix function relies on access to the swaprows and replacerow functions. While the swaprows function replaces three lines of R code wherever it is used, the replacerow function only replaces one. In a sense, it serves to help us read the code better, which is advantageous when the R code needs to be extended. It can also help if we wish to implement a related algorithm. In the case of `refmatrix`, it would be useful to extend the function to find the reduced row echelon form of a matrix.

Like row echelon form, the reduced row echelon form of the matrix is the result of a series of row operations that transform a matrix into the row-equivalent upper-triangular matrix. However, each row is scaled by the multiplicative inverse of the leading entry of the row. In other words, the row's entries are divided by the leading entry. All of the leading entries in the matrix should be 1 in a reduced row echelon matrix.

The implementation of function `rrefmatrix` is provided in Function 3.5. The implementation itself is almost identical to the implementation of Function `refmatrix`. There are only two differences. The first is a pair of lines, after any necessary row swap. These lines first capture the pivot value then divide the row by the pivot value. The second difference is the terms of the loop following row scaling. Instead of reducing each row after the current row, the modified `rrefmatrix` reduces each row, except the pivot row. This eliminates the entries above the current pivot, in addition to those below, as in `refmatrix`. The resulting matrix is in reduced row echelon form.

This is not the only implementation option available for reduced row echelon form. While this implementation reuses the underlying code from `refmatrix`, it does not use the function itself. It would be possible for `rrefmatrix` to begin by calling `refmatrix` and then scaling each row of the result. Such an implementation is likely slower in the general case, but the separation of duties inherent in such an implementation might outweigh the downsides of selecting this implementation strategy. It is incumbent on us, as implementors, to make that decision but also ensure that the decision can be changed with as little work as possible. In this case, there is no reason the implementation and design of `rrefmatrix` should be an issue for any of its users. Provided a matrix is given to the function, one should be returned in the form requested.

The function itself, behaves as expected using the same examples from before. For a square matrix, `rrefmatrix` should only return an identity matrix of the size of the input matrix,

```
> A <- matrix(c(5, 5, 5, 8, 2, 2, 6, 5, 4), 3)
> rrefmatrix(A)
     [,1] [,2] [,3]
[1,]    1    0    0
[2,]    0    1    0
[3,]    0    0    1
> A <- matrix(c(2, 4, 2, 4, 9, 4, 3, 6, 7, 7, 3, 9), 3)
> rrefmatrix(A)
     [,1] [,2] [,3]    [,4]
[1,]    1    0    0   24.75
[2,]    0    1    0  -11.00
[3,]    0    0    1    0.50
> A <- matrix(c(2, 8, 5, 5, 1, 2, 3, 8, 4), 3)
> rrefmatrix(A)
```

```
         [,1] [,2] [,3]
[1,]      1    0    0
[2,]      0    1    0
[3,]      0    0    1
```

Note that the last example which previously showed some numerical error in one row, is a clean matrix, without evidence of numerical error. In getting to row echelon form, the value in the middle of the matrix was 19, which became a divisor for a row replacement on row 3. Dividing something by 19 will not yield an exact result in floating point arithmetic and the error creeps in as something close to but not quite zero. In getting to reduced row echelon form, this is not a problem, in this case. Our row scaling happens before establishing any row replacement divisors, and the pivot values are always 1. And even the oddities of floating point arithmetic can handle dividing by 1. However, the 19 became a divisor for the third column of the middle row, so we did not earn a free lunch, merely shifted who paid and when.

We want to be able to reduce matrices as it is the first method we have for solving linear equations. Given a set of linear equations,

$$2x_1 + x_2 - x_3 = 1 \tag{3.8}$$

$$3x_1 + 2x_2 - 2x_3 = 1 \tag{3.9}$$

$$x_1 - 5x_2 + 4x_3 = 3 \tag{3.10}$$

we are interested in finding a set of values for x_1, x_2, and x_3 that solve the equations. From linear algebra, we know we can define a matrix A that includes the coefficient of the equation and a vector $\mathbf{b}$ that is the right-hand side of the equations so that $A\mathbf{x} = \mathbf{b}$. Further, we know we can solve for x by creating a matrix of the form $[A\,\mathbf{b}]$, and, through row operations, finding the reduced row echelon form. Then the far right-hand column will have the solution x represented as a vector. We can use `rrefmatrix` to demonstrate this:

```
> (A <- matrix(c(2, 3, 1, 1, 2, -5, -1, -2, 4), 3))
       [,1] [,2] [,3]
[1,]    2    1   -1
[2,]    3    2   -2
[3,]    1   -5    4
> (b <- c(1, 1, 3))
[1] 1 1 3
> rrefmatrix(cbind(A, b))
                b
[1,] 1 0 0 1
[2,] 0 1 0 2
[3,] 0 0 1 3
```

The solution is embedded in the final column, and is a familiar set of numbers. Because of how often we might do this, we could use a utility function for convenience to solve sets of linear equations and return the results. This

```
solvematrix <- function(A, b) {

    m <- cbind(A, b)
    m <- rrefmatrix(m)
    x <- m[, ncol(m)]

    return(x)
}
```

R Function 3.6
Use reduced row echelon form to solve linear equations

function, given in Function 3.6, does just this. It accepts two arguments, A and b, and uses `rrefmatrix` to solve for x, which it returns. Barring exceptional circumstances, we are better off using the internal function for solving. However, there is no comparable function for finding either a row echelon form or reduced row echelon form of a matrix.

The example results are,

```
> solvematrix(A, b)
[1] 1 2 3
```

which returns the right-most column of the results of the reduced row echelon matrix. Generally, Gaussian elimination is not the best approach for solving matrix equations. Gaussian elimination requires a large number of steps and for large matrices, can take considerable time. These implementations are slower still for being in interpreted R code. R, itself, provides an alternative. The `solve` function is built in to R and accepts two parameters, A and b, to find x:

```
> solve(A, b)
[1] 1 2 3
```

The built-in function is safer in that it provides error-checking and other support necessary to ensure proper function of the algorithm. In addition, the function does not solve over- or underdetermined systems; the matrix A must be square. The underlying implementation of `solve` uses the high speed and very powerful LAPACK library, which is the basis for a number of numerical and matrix equation packages.

3.2.2   Tridiagonal Matrices

There are a number of specialized matrix forms that can lead to less complicated solvers. Some of these center around banded matrices. Banded matrices generally consist of diagonal entries, where entries not on the diagonals are zero. The simplest case is a diagonal matrix such that all of the entries on the main diagonal are nonzero and all other entries are zero. For instance, consider the n-by-n matrix A such that,

$$A = \begin{bmatrix} d_1 & 0 & \cdots & 0 \\ 0 & d_2 & \cdots & 0 \\ \vdots & \vdots & \ddots & \vdots \\ 0 & 0 & \cdots & d_m \end{bmatrix}.$$

(3.11)

Assuming that the relationship $A\mathbf{x} = \mathbf{b}$ holds for some vectors $\mathbf{x}$ and $\mathbf{b}$, then the augmented matrix $[A\mathbf{b}]$ will be zero for all entries except for the far right-hand column and the original main diagonal of A. This is equal to the system of equations,

$$d_1 x_1 + 0 x_2 + \cdots + 0 x_n = b_1 \tag{3.12}$$
$$0 x_1 + d_2 x_2 + \cdots + 0 x_n = b_2 \tag{3.13}$$
$$\vdots$$
$$0 x_1 + 0 x_2 + \cdots + d_n x_n = b_n. \tag{3.14}$$

That system of equations, after eliminating all of the zero terms, is a system of equations such that $d_i x_i = b_i$ where i is some integer from 1 to n. And that is trivial to solve for $x_i = b_i/d_i$, since the values of both b_i and d_i are known. In the terms of linear algebra, this is the final reduction step for each row, dividing the row by the leading term. This process will not succeed if any of the entries off the main diagonal are non-zero.

Based on this intuitive understanding of the simplest band matrix above, we can extend the logic to handle tridiagonal matrices. Tridiagonal matrices are those matrices where there are three bands of entries consisting of the main diagonal and the diagonals immediately adjacent to the main diagonal. All other entries must be zero. For example, suppose there is a matrix A such that,

$$A = \begin{bmatrix} d_1 & u_1 & 0 & 0 & 0 \\ l_2 & d_2 & u_2 & 0 & 0 \\ 0 & l_3 & \ddots & \ddots & 0 \\ 0 & 0 & \ddots & \ddots & u_{n-1} \\ 0 & 0 & 0 & l_n & d_n \end{bmatrix},$$

(3.15)

In the single-band diagonal matrix, the matrix can be viewed as the result of a nearly completed Gaussian elimination. The same is true of a tridiagonal matrix, except it is not nearly as complete. But it is pretty far along and we can take advantage of that in two ways.

The first advantage is the efficiency of the row reduction process. We will first assume there is a value $\mathbf{x}$ for which $A\mathbf{x} = \mathbf{b}$, with known $\mathbf{b}$, and we are attempting to solve the linear equations using the augmented matrix $[A\mathbf{b}]$. The key here is to recognize that for any non-zero entry in the matrix, there is at

```
tridiagmatrix <- function(L, D, U, b) {
    n <- length(D)
    L <- c(NA, L)

    ##   The forward sweep
    U[1] <- U[1] / D[1]
    b[1] <- b[1] / D[1]
    for(i in 2:(n - 1)) {
        U[i] <- U[i] / (D[i] - L[i] * U[i - 1])
        b[i] <- (b[i] - L[i] * b[i - 1]) /
            (D[i] - L[i] * U[i - 1])
    }
    b[n] <- (b[n] - L[n] * b[n - 1]) /
        (D[n] - L[n] * U[n - 1])

    ##   The backward sweep
    x <- rep.int(0, n)
    x[n] <- b[n]
    for(i in (n - 1):1)
        x[i] <- b[i] - U[i] * x[i + 1]

    return(x)
}
```

R Function 3.7
Tridiagonal matrix algorithm

most one entry above it and one below it to consider in the Gaussian elimination process. In addition, because of the structure of a tridiagonal matrix, the matrix is essentially prepivoted, from the viewpoint of Gaussian elimination.

The process here is in two parts with a "forward sweep" and a "backward sweep." During the forward sweep, each element of the l band shown in Equation 3.15 will be eliminated by row reduction. During the backwards sweep, each element of the u band will be eliminated. But because so many of the results involve multiplying something by zero, which results in no change, it isn't necessary to carry out the row reduction across the entire row. Just on the elements that will be affected. The tridiagonal matrix algorithm, tridiagmatrix, is implemented in Function 3.7.

The second advantage, as seen from the implementation, is in memory storage. To store an m-by-n matrix requires mn storage locations. For a square matrix of size m, this grows exponentially. When so many entries of the matrix are zero, and there is a predictable pattern of zeros in the matrix, we can store the matrix using less space. The tridiagmatrix function only stores the entries on the three diagonals. For a matrix of size m, that requires $3m - 2$ storage locations, which is less than or equal to m^2 for any $m \geq 2$, which is also the minimum size of any matrix. Another advantage to the tridiagonal matrix algorithm is that certain intermediate values of the matrix, such as the zeroed entries below the main diagonal, are never used again, so they are not stored. At the end of Function 3.7, the values of l and d are unchanged.

Consider the matrix equation,

$$A = \begin{bmatrix} 3 & 4 & 0 & 0 \\ 4 & 5 & 2 & 0 \\ 0 & 2 & 5 & 3 \\ 0 & 0 & 3 & 5 \end{bmatrix} \quad \mathbf{x} = \begin{bmatrix} 20 \\ 28 \\ 18 \\ 18 \end{bmatrix}. \tag{3.16}$$

Solving equation 3.16 via row reduction is not particularly difficult, even if solved by hand. However, the advantages from both a storage and computational standpoint grow larger with larger matrices and should be taken if possible. The solution to 3.16 is,

```
> l <- u <- c(4, 2, 3); d <- c(3, 5, 5, 5)
> b <- c(20, 28, 18, 18)
> tridiagmatrix(l, d, u, b)
[1] 4 2 1 3
```

For matrices with more than three bands, it is possible to construct more complicated algorithms for solving them by assuming the matrix is partially reduced. However, these yield fewer and fewer advantages over generalized algorithms as the number of bands increases, but gain advantages, with respect to both storage and processing efficiency, as the overall size of the matrix increases. When deploying matrix algorithms, care should be taken to ensure the tradeoffs balance, and the chosen implementation fits the use case.

3.3   Matrix Decomposition

Those solutions to banded matrices presume a shortcut through the solution. The presupposition is that the work is already done and there is not much left to do but finish the job. Often, we are not so lucky that the groundwork is done for us and we have to start from scratch. It would be better, in a sense, if we could preclear a path and have a shortcut available to us. That is what matrix factorization algorithms do. By preworking a matrix, the factorization gets us closer to a solution than when we started, and this helps when we must repeatedly process a linear system for different values.

3.3.1   LU Decomposition

LU decomposition offers a better way of solving systems of equations using matrices. LU decomposition is named for the two components of the decomposition, a lower triangular matrix and an upper triangular matrix. The matrices L and U are selected so that $A = LU$. This turns out to be a powerful tool because,

$$Ax = b \tag{3.17}$$

$$L(Ux) = b. \tag{3.18}$$

Since U is a matrix of m rows and x is a vector of m elements, the result of their multiplication is also a vector of m elements. We will let that temporary vector be called t such that

$$t = Ux, \tag{3.19}$$

and then, substituting t for Ux,

$$Lt = b. \tag{3.20}$$

From this initial assessment, it is apparent that we can solve for x by initially solving equation 3.20 for t, then solving equation 3.19 for x, yielding the result we were looking for.

We can solve equations 3.20 and 3.19 using any matrix solving algorithm. However, because L is a lower triangular matrix, with all zero entries above the main diagonal, solving for t takes relatively few steps. This, like the tridiagonal matrices, takes advantage of a shortcut, but this time one we create to ease computation. The lower diagonal matrix L will be a square matrix of size m where m is the number of rows in the original matrix A. Accordingly, equation 3.20 will have a form like,

$$Lt = \begin{bmatrix} 1 & 0 & \cdots & 0 \\ l_{2,1} & 1 & \cdots & 0 \\ \vdots & \vdots & \ddots & \vdots \\ l_{m,1} & l_{m,2} & \cdots & 1 \end{bmatrix} \begin{bmatrix} t_1 \\ t_2 \\ \vdots \\ t_m \end{bmatrix} = \begin{bmatrix} b_1 \\ b_2 \\ \vdots \\ b_m \end{bmatrix}. \tag{3.21}$$

In general, a lower triangular matrix will not necessarily have only ones on the main diagonal, but the process of LU decomposition leads to it here.

Remembering that equation 3.21 is a system of equations, it is obvious that $t_1 = b_1$. Further, we can substitute this value into the second row and determine that $t_2 = b_2 - l_{2,1}t_1$. Further, we can continue this process through the entire matrix. The generalized solution for t_i is,

$$t_i = b_i - \sum_{j=1}^{i-1} l_{i,j}t_i. \tag{3.22}$$

This process is called *forward substitution*. Similarly, with t resolved, we can solve equation 3.19,

$$Ux = \begin{bmatrix} u_{1,1} & u_{1,2} & \cdots & u_{1,n} \\ 0 & u_{2,2} & \cdots & u_{2,n} \\ \vdots & \vdots & \ddots & \vdots \\ 0 & 0 & \cdots & u_{m,n} \end{bmatrix} \begin{bmatrix} x_1 \\ x_2 \\ \vdots \\ x_m \end{bmatrix} = \begin{bmatrix} t_1 \\ t_2 \\ \vdots \\ t_m \end{bmatrix}. \tag{3.23}$$

There are some subtle differences between these two solutions. First, the matrix U does not necessarily have ones on the main diagonal. Second, it is not necessarily square, though if A is square, so is U, and square matrices are the ones we are interested in solving. Like in the process of forward substitution, we can note that $x_m = t_m/u_{m,n}$, and so forth from the last row to the first row of the matrix. This process is called "backward substitution."

Finding the LU decomposition is based on row operations. First, we must find an upper triangular matrix corresponding to A. There are infinitely many solutions, but the simplest to obtain is the row echelon form of the matrix, just like in Function 3.4. Second, L should be a lower triangular matrix that reduces to $\mathbf{I}$ by following the same row operations that produces U. We can use the Doolittle algorithm to generate L, where the value of each entry in the lower triangular matrix is the multiplier used to eliminate the corresponding entry for each row replacement. This algorithm is given in `lumatrix` given in Function 3.8.

In practice, the Gaussian elimination process used to obtain U may encounter a zero in the pivot column, necessitating a row swap for a non-zero pivot. If this happens, then A may be row equivalent to LU, but not identical. The implementation of `lumatrix` preserves this information by returning three matrices, instead of two. The third matrix, P, initially holds an identity matrix of size m, but if a row swap is necessary, the same swap is executed on P. Therefore, in practice $A = PLU$ and the multiplication by P restores the order of the rows.

We can see the process in action and verify the results in this example.

```
> (A <- matrix(c(0, 1, 7, 1, 5, -1, -2, 9, -5), 3))
     [,1] [,2] [,3]
[1,]    0    1   -2
[2,]    1    5    9
[3,]    7   -1   -5
> (decomp <- lumatrix(A))
$P
     [,1] [,2] [,3]
[1,]    0    1    0
[2,]    1    0    0
[3,]    0    0    1

$L
     [,1] [,2] [,3]
[1,]    1    0    0
[2,]    0    1    0
[3,]    7  -36    1

$U
     [,1] [,2] [,3]
[1,]    1    5    9
```

```
lumatrix <- function(m) {
    count.rows <- nrow(m)
    count.cols <- ncol(m)
    piv <- 1

    P <- L <- diag(count.cols)
    for(row.curr in 1:count.rows) {
        if(piv <= count.cols) {
            i <- row.curr
            while(m[i, piv] == 0 && i < count.rows) {
                i <- i + 1
                if(i > count.rows) {
                    i <- row.curr
                    piv <- piv + 1
                    if(piv > count.cols)
                        return(list(P = P, L = L, U = m))
                }
            }
            if(i != row.curr) {
                m <- swaprows(m, i, row.curr)
                P <- swaprows(P, i, row.curr)
            }
            for(j in row.curr:count.rows)
                if(j != row.curr) {
                    k <- m[j, piv] / m[row.curr, piv]
                    m <- replacerow(m, row.curr, j, -k)
                    L[j, piv] <- k
                }
            piv <- piv + 1
        }
    }

    return(list(P = P, L = L, U = m))
}
```

R Function 3.8
The matrix LU decomposition function

```
[2,]    0    1   -2
[3,]    0    0 -140
> decomp$P %*% decomp$L %*% decomp$U
     [,1] [,2] [,3]
[1,]    0    1   -2
[2,]    1    5    9
[3,]    7   -1   -5
```

The end result is indeed the value *A* created at the beginning. In addition, because the value of $a_{1,1}$ is 0, `lumatrix` was forced to swap the first and second rows, and this is reflected in the value of P. From an implementation standpoint, there are three matrices to return, so the function creates them as a list. Each element of the list is addressed using the dollar sign operator. Another potential option is to create a class representing LU decompositions.

The number of operations necessary to produce the LU decomposition is the same as the number to transform a matrix to row echelon form. In a sense, the total cost of production is limited to the cost producing U. The swaps necessary to generate P are considered costless and the generation of L comes from storing the values used to generate U. Therefore, we can consider the numerical cost of both P and L to be free.

Despite the relatively low numerical cost, the true advantage comes from repetition. Given a problem where the equation $A\mathbf{x} = \mathbf{b}$ must be solved for many different values of $\mathbf{b}$, the decomposition can be executed once and forward and backwards substitution used to quickly solve for values of $\mathbf{t}$ and $\mathbf{x}$.

A second numerical advantage exists that unnecessary operations are not performed in the reduction to reduced row echelon form. This allows the matrix to miss some pitfalls where, through cascading errors, slightly numerically incorrect answers can become big errors, though not all pitfalls are avoided.

Because of these advantages, many computer-based matrix solvers use LU decomposition internally, rather than row operations to reduced row echelon form. In addition, R provides an implementation through the `Matrix` library in the `lu` function. That implementation is well developed and the package also includes methods for handling triangular matrices and many other specialized matrix types.

3.3.2   Cholesky Decomposition

The Cholesky decomposition of a matrix provides an alternative matrix factorization such that $A = LL^*$ where L^* is the conjugate transpose of the matrix L. In our case, we are only working with real matrices with real values and the imaginary parts are zero. So for our purposes here, the matrix L^* is just the transpose of the matrix L.

Like the LU decomposition, the Cholesky decomposition can be used to solve matrix equations. Further, finding the Cholesky decomposition is notably faster than the LU decomposition. However, it is more limited. The Cholesky decomposition can only be used on symmetric positive definite matrices. Symmetric matrices are matrices that are symmetric about the main diagonal; mathematically, for all i and j, $a_{i,j} = a_{j,i}$, for a matrix A. Positive definite means that each of the pivot entries is positive. In addition, for a positive definite matrix, the relationship $\mathbf{x}A\mathbf{x} > 0$ for all vectors, $\mathbf{x}$. This has applications for curve fitting and least squares approximations.

Because L^* is the transpose of L, then $l^*_{i,j} = l_{j,i}$ for all values of i and j. Without this constraint, the LU decomposition is quite similar. But with this constraint, the values of both L and L^* have to be carefully selected so that the relationship $A = LL^*$ holds. For an m-by-m matrix,

$$\begin{bmatrix} a_{1,1} & a_{1,2} & \cdots & a_{1,m} \\ a_{2,1} & a_{2,2} & \cdots & a_{2,m} \\ \vdots & \vdots & \ddots & \vdots \\ a_{m,1} & a_{m,2} & \cdots & a_{m,n} \end{bmatrix} = \begin{bmatrix} l_{1,1} & 0 & \cdots & 0 \\ l_{2,1} & l_{2,2} & \cdots & 0 \\ \vdots & \vdots & \ddots & \vdots \\ l_{m,1} & l_{m,2} & \cdots & l_{m,m} \end{bmatrix} \begin{bmatrix} l_{1,1} & l_{2,1} & \cdots & l_{m,1} \\ 0 & l_{2,2} & \cdots & l_{m,2} \\ \vdots & \vdots & \ddots & \vdots \\ 0 & 0 & \cdots & l_{m,m} \end{bmatrix}. \quad (3.24)$$

For each entry of the matrix A,

$$a_{i,j} = \sum_{k=1}^{m} L_{i,k} L_{k,j}^*. \quad (3.25)$$

According to equation 3.24, many of the values of $L_{i,k}$ and $L_{k,j}^*$ are zero. Breaking these down, the result is that,

$$l_{i,i} = \sqrt{\left(a_{i,i} - \sum_{k=1}^{i-1} l_{i,k}^2 \right)}, \quad (3.26)$$

defines the entries on the main diagonal. The entries off the diagonal are defined as,

$$l_{i,j} = \frac{1}{l_{i,i}} \left(a_{i,j} - \sum_{k=1}^{i-1} l_{i,k} l_{j,k} \right). \quad (3.27)$$

The implementation of equations 3.26 and 3.27 are given in Function 3.9, the `choleskymatrix` function. The result of `choleskymatrix` is actually the value of L^*, the upper triangular matrix. To receive L, use the `t` transpose function in R.

```
> (A <- matrix(c(5, 1, 2, 1, 9, 3, 2, 3, 7), 3))
     [,1] [,2] [,3]
[1,]    5    1    2
[2,]    1    9    3
[3,]    2    3    7
> (L <- choleskymatrix(A))
         [,1]      [,2]      [,3]
[1,] 2.236068 0.4472136 0.8944272
[2,] 0.000000 2.9664794 0.8764598
[3,] 0.000000 0.0000000 2.3306261
> t(L) %*% L
     [,1] [,2] [,3]
[1,]    5    1    2
[2,]    1    9    3
[3,]    2    3    7
```

```
choleskymatrix <- function(m) {
    count.rows <- nrow(m)
    count.cols <- ncol(m)

    L = diag(0, count.rows)
    for(i in 1:count.rows) {
        for(k in 1:i) {
            p.sum <- 0
            for(j in 1:k)
                p.sum <- p.sum + L[j, i] * L[j, k]
            if(i == k)
                L[k, i] <- sqrt(m[i, i] - p.sum)
            else
                L[k, i] <- (m[k, i] - p.sum) / L[k, k]
        }
    }
    return(L)
}
```

R Function 3.9
The Cholesky decomposition

After `choleskymatrix` has returned a value of L^*, it and the transpose can be used just like the upper and lower triangular matrices in the LU decomposition.

R includes a built-in function for this, called `chol`. This function returns the upper triangular matrix, and this is why the design choice was made to return the upper triangular matrix in `choleskymatrix`. In addition, there is an implementation of the Cholesky decomposition in the previously mentioned `Matrix` package that complements the matrix objects in `Matrix`.

3.4   Iterative Methods

To now, we have looked at three methods of solving matrix equations based on row operations. Even the Cholesky decomposition, which finds the individual values of elements, is based on the multiplicative relationship of the underlying matrices, just like the LU decomposition. Now we are interested in a very different approach, based on iterative methods.

Iterative methods start with an estimate of the final value. After applying some treatment to the estimate, an updated estimate is returned. This updated estimate is again subject to the treatment, and the process continues by restarting with the updated estimate. The process completes when some

```
vecnorm <- function(b) {
    return(sqrt(sum(b^2)))
}
```

R Function 3.10
Find the norm of a vector

threshold is met. That threshold may be that the difference between two iterations is so small that the point of diminishing returns is met. Or the threshold may be based on not exceeding some fixed number of iterations.

In practice, we will use both thresholds. In principle, we will want to see our methods reach the point of diminishing returns. However, we will use a fixed limit of iterations as a sort of "safety check" to ensure the algorithm does not run away. When the maximum iteration count is reached, the current estimate of the result is returned, regardless of its estimated quality.

As part of this, we need a method to check the distance between the current iteration and the last iteration. We could examine the distance between the iteration of each element and ensure that none exceed some threshold. Another approach is to use the absolute distance between the two iterations' vectors. In the general case, the threshold is that

$$||\mathbf{x}^{(n+1)} - \mathbf{x}^{(n)}|| < t_0, \tag{3.28}$$

where $\mathbf{x}^{(n)}$ is the nth iteration of the algorithm and t_0 is the maximum acceptable tolerance.

Function 3.10 provides a one-liner that implements the vector norm function, vecnorm. We use this because the norm function within R, which we might expect to calculate the vector norm, instead calculates the matrix norm. While related, the functions are different and the norm function does not accept an array parameter, anyway. The function only accepts a single vector as a parameter:

```
> (x <- c(4, 8, 7, 2))
[1] 4 8 7 2
> vecnorm(x)
[1] 11.53256
```

Like row operations, this is provided for both semantic and programming convenience. This function will be used in the check tolerance step for each of the iterative matrix solvers following.

3.4.1   Jacobi Iteration

For solving matrices iteratively, we can begin, again, with the premise there is a matrix A and vectors $\mathbf{x}$ and $\mathbf{b}$, so that $A\mathbf{x} = \mathbf{b}$, which should be a familiar relationship, at this point. Using Jacobi's method, we can first observe there are matrices R and D that exist such that $A = R + D$. In fact, an infinite number of pairs, R and D, exist that meet that requirement. Following the rules for linear algebra and matrix addition, we observe,

$$Ax = b \tag{3.29}$$
$$Rx + Dx = b \tag{3.30}$$
$$Dx = b - Rx \tag{3.31}$$
$$x = D^{-1}(b - Rx) \tag{3.32}$$

Which is nearly a solution for **x**. If we rewrite equation 3.32 so that

$$x^{(n+1)} = D^{-1}(b - Rx^{(n)}) \tag{3.33}$$

it is clear that, at least in some cases, the iteration will converge to some value, **x**.

Like other methods of solving linear equations, Jacobi's method relies on preworking part of the problem to ensure a quicker solution. In this case, we find the inverse of the matrix quickly by deciding that the matrix D should be a strictly diagonal matrix with entries corresponding to the main diagonal of A. A strictly diagonal matrix is trivial to inverse, as its inverse is a diagonal matrix equal to one divided by the corresponding element of the original matrix. The matrix R is identical to the matrix A, except the main diagonal is only zeros. For example,

```
> (A <- matrix(c(5, 2, 1, 2, 7, 3, 3, 4, 8), 3))
     [,1] [,2] [,3]
[1,]    5    2    3
[2,]    2    7    4
[3,]    1    3    8
> (D <- diag(1 / diag(A)))
     [,1]      [,2]  [,3]
[1,]  0.2 0.0000000 0.000
[2,]  0.0 0.1428571 0.000
[3,]  0.0 0.0000000 0.125
> (R <- A - diag(diag(A)))
     [,1] [,2] [,3]
[1,]    0    2    3
[2,]    2    0    4
[3,]    1    3    0
```

As $R = A - D$, so $D + R = A$. Further, after finding the inverse of D, the iteration in equation 3.33 can be repeated without conducting any row operations, Gaussian elimination, or other computationally expensive steps. The most complex steps taken are multiplying a matrix by a vector. And the function jacobi, given in Function 3.11 implements this iteration.

There are some requirements to complete the iteration successfully, Most

```
jacobi <- function(A, b, tol = 10e-7, maxiter = 100) {
    n <- length(b)
    iter <- 0

    Dinv <- diag(1 / diag(A))
    R <- A - diag(diag(A))
    x <- rep(0, n)
    newx <- rep(tol, n)

    while(vecnorm(newx - x) > tol) {
        if(maxiter > iter) {
            warning("iterations maximum exceeded")
            break
        }
        x <- newx
        newx <- Dinv %*% (b - R %*% x)
        iter <- iter + 1
    }

    return(as.vector(newx))
}
```

R Function 3.11
Jacobi iteration for diagonally dominant matrices

importantly, the matrix must be diagonally dominant. That means the absolute values of the diagonal entries must be larger than the sum of the absolute value of all other entries in a given column. And none of those values can be zero, or the inverse of D will fail, due to a division error. If these properties hold, the Jacobi iteration will successfully complete. Also, even if these properties do not hold for the given matrix, the Jacobi iteration may still converge, but there is no guarantee.

Using A defined above,

```
> (b <- c(40, 39, 55))
[1] 40 39 55
> jacobi(A, b)
[1] 1e-06 1e-06 1e-06
```

There are two unused options to the `jacobi` function. First, there is an option `tol` that defines the tolerance. It is the largest value for which $||\mathbf{x}^{(n+1)} - \mathbf{x}^{(n)}||$ is acceptable. That value defaults to 1×10^{-6}, or one part per million. That's a fairly small value and given that the true solution to this example,

```
> solvematrix(A, b)
[1] 4 1 6
```

The results are very good.

The second constraint is `maxiter`, which defines the maximum number of iterations the algorithm will test. If the value of $\mathbf{x}^{(n)}$ has not converged within an acceptable range by then, the function will still terminate and return the current value of $\mathbf{x}$. This implementation leaves some details to be desired, such as knowing whether or not the function has returned because the maximum number of iterations has been exceeded or because the converge threshold has been reached.

We can use `maxiter` to trace the convergence process in ten-step increments.

```
> jacobi(A, b, maxiter = 10)
[1] 1e-06 1e-06 1e-06
> jacobi(A, b, maxiter = 20)
[1] 1e-06 1e-06 1e-06
> jacobi(A, b, maxiter = 30)
[1] 1e-06 1e-06 1e-06
> jacobi(A, b, maxiter = 40)
[1] 1e-06 1e-06 1e-06
> jacobi(A, b, maxiter = 50)
[1] 1e-06 1e-06 1e-06
```

While easy to calculate, a large number of steps, relatively speaking, can be necessary to find a solution. Even after 50 iterations, the estimate is still not within the default threshold level.

The last design decision included here is the selection of the initial guess. To simplify the loop, the last guess is set to all zeros. The next guess is a vector composed of the tolerance times 10. This ensures that the first time through the loop, the norm of the difference will exceed the tolerance. After this loop test is completed, the vector of zeros is dropped and the initial starting guess is that vector of the tolerances. The initial guess can be fairly arbitrary and selecting a value too far from the correct value of $\mathbf{x}$ will increase the iterations required to converge. However, the process will still converge.

3.4.2 Gauss–Seidel Iteration

Given the parameters laid out in equation 3.32, it is reasonable to assume that other matrices A_α and A_β can be used in place of R and D, provided $A_\alpha + A_\beta = A$. In general, this assumption is both correct and useful, subject to certain provisos. First, there is an ongoing condition that the original matrix A be of some constrained form, such as diagonally dominant. Second, the assumption is only useful if the addends A_α and A_β can be found relatively easily.

The Gauss–Seidel iterative method decomposes the matrix into an upper triangular matrix, U and a lower triangular matrix, L. These are not the same upper and lower triangular matrices that form the LU decomposition. With Gauss–Seidel, the upper triangular matrix is the entries of the matrix A above

```
gaussseidel <- function(A, b, tol = 10e-7, maxiter = 100) {
    n <- length(b)
    iter <- 0

    L <- U <- A
    L[upper.tri(A, diag = FALSE)] <- 0
    U[lower.tri(A, diag = TRUE)] <- 0
    Linv <- solve(L)

    x <- rep(0, n)
    newx <- rep(tol * 10, n)

    while(vecnorm(newx - x) > tol) {
        if(maxiter > iter) {
            warning("iterations maximum exceeded")
            break
        }
        x <- newx
        newx <- Linv %*% (b - U %*% x)
        iter <- iter + 1
    }

    return(as.vector(newx))
}
```

R Function 3.12
Gauss–Seidel iteration for diagonally dominant matrices

the main diagonal. The lower triangular matrix is the entries of the original matrix below the main diagonal and including the main diagonal. Like with the Jacobi iterative process, the undefined entries of both U and L are set to zero. The iterated equation for Gauss–Seidel is conceptually identical to the equation for Jacobi's method, so that,

$$\mathbf{x}^{(n+1)} = L^{-1}(\mathbf{b} - U\mathbf{x}^{(n)}). \tag{3.34}$$

Aside from the construction of the matrices L and U for the Gauss–Seidel iteration, the iterative process itself is identical to the process in equation 3.33. The function gaussseidel and jacobi are identical except for the variable names used. The implementation uses the R internals for creating upper and lower triangular matrices to create the matrices for L and U and the inverse of the matrix L is found using the solve function, available in R.

It is worth noting this implementation adopts, where feasible, the same design decisions as the implementation of jacobi. In particular, the default values of the tolerance and maximum iteration are the same. In addition, the tolerance calculation and the initial guess at the value of **x** are also the same as in jacobi. These assumptions are based on consistency and are not necessarily reflective of how either of these iterative processes operates.

Because of this consistency, the operation of the function is the same with the values of A and **b** passed in. Using the example of A created for the Jacobi iteration example, we can see,

```
> gaussseidel(A, b)
[1] 1e-05 1e-05 1e-05
```

Of note, using five-step iterations, we can observe that the correct value is created more quickly than using the Jacobi method:

```
> gaussseidel(A, b, maxiter = 5)
[1] 1e-05 1e-05 1e-05
> gaussseidel(A, b, maxiter = 10)
[1] 1e-05 1e-05 1e-05
> gaussseidel(A, b, maxiter = 15)
[1] 1e-05 1e-05 1e-05
> gaussseidel(A, b, maxiter = 20)
[1] 1e-05 1e-05 1e-05
```

This accelerated convergence comes at the cost of applicability. The Gauss–Seidel iteration is only guaranteed to converge if the matrix A is diagonally dominant or symmetric and positive definite. This limits the applications of the algorithm to certain subsets of matrices, though those matrices are not uncommon.

If the matrix does not meet these requirements, then failure to converge can be somewhat spectacular. Using the same value of **b** and a random matrix M,

```
> (M <- matrix(c(-9, 4, -5, -3, -7, 8, -4, 7, 5), 3))
      [,1] [,2] [,3]
[1,]   -9   -3   -4
[2,]    4   -7    7
[3,]   -5    8    5
> solve(M, b)
[1] -7.026104 -2.158635  7.427711
```

Based on the results of the solve function, the solution vector has no element with an absolute value greater than ten. But the gaussseidel function grows steadily out of control. Based on five-step iterations,

```
> gaussseidel(M, b, maxiter = 5)
[1] 1e-05 1e-05 1e-05
> gaussseidel(M, b, maxiter = 30)
[1] 1e-05 1e-05 1e-05
> gaussseidel(M, b, maxiter = 50)
[1] 1e-05 1e-05 1e-05
> gaussseidel(M, b, maxiter = 70)
```

```
[1] 1e-05 1e-05 1e-05
> gaussseidel(M, b, maxiter = 100)
[1] 1e-05 1e-05 1e-05
```

That number will continue to grow until the computer runs out of storage.

Both the Jacobi and Gauss–Seidel iterative methods provide some advantages over other methods for solving a matrix. First, neither requires finding the matrix inverse, a quite computationally intensive task. Second, in practice, many matrix equations involve very sparse matrices, those with a large number of zero entries. With sparse matrices, the repeated multiplicative step can be handled quickly and with a higher degree of accuracy than the associated matrix inversions. Certainly, for large matrix equations, fewer operations are necessary for a good approximation of the final result with iterative methods.

3.5 Applications

There are numerous applications of linear algebra and solutions to matrices ranging from engineering, to physics, to statistical analysis. It might be simpler to list disciplines not using linear algebra, implicitly or explicitly. Because of this, we will see a number of applications throughout this book going forward.

But before we get there, we want to look at the least squares, which is an application R users are most familiar with.

3.5.1 Least Squares

In some ways, least squares is the defining application of linear algebra. Given a set of data points, we might want to find the line that "best fits" the points. Least squares is one answer to that problem. The line found by least squares is the line that minimizes the distance between the line and the data points. It may not exactly fit the points. Unless three points are precisely colinear, a single line cannot fit all three. However, a unique line exists that has the smallest combined distance between each of the three points and the line, which will be somewhere nearby.

To find this line, we will begin with a set of data points. Built into R is a dataset called **trees** that can be accessed by that name. The dataset contains 31 rows, which is inconvenient to print; so we use the **head** function to provide the first six entries, to get a flavor of the dataset.

```
> head(trees)
  Girth Height Volume
1  8.3     70   10.3
2  8.6     65   10.3
3  8.8     63   10.2
```

```
4   10.5      72    16.4
5   10.7      81    18.8
6   10.8      83    19.7
```

Girth is the diameter of a given tree in inches; height is the height in feet; volume is the amount of timber in cubic feet obtained from the tree. Logically, we might assume that the volume is a function of both the girth and height of the tree. We might define the relationship by the equation,

$$\text{volume} = \beta_{\text{girth}}\text{girth} + \beta_{\text{height}}\text{height} + \beta_0, \qquad (3.35)$$

where β_0 is the y-intercept of the volume line and the other values of β are coefficients of the girth and height. Volume is called the "response variable," and might be stated as y in a general case. Girth and height, called x_n in the general case, are the input variables. Least squares allows us to test that relationship.

Given the large number of observations, relative to the number of input variables, finding a perfect fit line is probably out of the question. Least squares, however, finds just the coefficients, the values of β, that create a best fit line. That best fit is defined as the smallest sum of the squares of the distances from the observation points to the best fit line. Hence, we have the name "least squares." The basis of least squares is the relatively simple equation,

$$A^{\mathsf{T}}Ax = A^{\mathsf{T}}b. \qquad (3.36)$$

In this equation, **b** is the result, or response variable, and A is a matrix containing the observation data. Solving this for x yields the values of β. To apply this measure, we will convert our dataset to appropriate matrices. For the trees, the value of **b** is the trees' volume measurements. The values of the A matrix are three columns, one containing the trees' girth measurements, one the height measurements, and one a column of ones.

```
> head(b <- trees$Volume)
[1] 10.3 10.3 10.2 16.4 18.8 19.7
> head(A <- cbind(1, trees$Girth, trees$Height))
     [,1] [,2] [,3]
[1,]    1  8.3   70
[2,]    1  8.6   65
[3,]    1  8.8   63
[4,]    1 10.5   72
[5,]    1 10.7   81
[6,]    1 10.8   83
```

Again, we have used the function **head** to produce the leading entries of the vector **b** and matrix A. Also, we have added a column to the matrix consisting of only the value 1, repeated. This allows the results to include that y-intercept. Otherwise, the columns are just the values of the girth and height. From here, we use the function **solvematrix**, our function that uses row reduction to solve matrix equations, to find the value of **b** from equation 3.36.

```
> (x <- solvematrix(t(A) %*% A, t(A) %*% b))
[1] -57.9876589    4.7081605    0.3392512
```

Each entry in x is the corresponding coefficient for each column in the matrix A. Our linear model would be,

$$\text{volume} = -57.988 + 4.708 \times \text{girth} + 0.3393 \times \text{height}. \qquad (3.37)$$

R provides, as part of its core statistical functions, a function for creating linear models, lm. And it would be the preferred tool for this example, as it is more flexible and handles the matrix algebra for us:

```
> lm(Volume ~ Girth + Height, data = trees)
Call:
lm(formula = Volume ~ Girth + Height, data = trees)

Coefficients:
(Intercept)        Girth        Height
  -57.9877       4.7082        0.3393
```

And lm provides the same results![2]

This process, given in equation 3.36 generalizes to any number of input variables and data points, limited only by the processing power and memory of the computer we are using for calculations. Because of this powerful generality, the least squares equation, and its derivatives, has become the cornerstone of statistical analysis. Least squares dominates statistical analysis especially in the social sciences, such as economics and sociology. Further, it is used to test complex interactions in medicine and biology and statistical mechanics.

Comments

Linear algebra underlies much of numerical analysis. In some cases, that's explicit, as we will see with interpolation in the next chapter. In other cases, it is implicit, as we will see with integration in the chapter after that. Having a sound toolkit for managing matrices, working with their contents, and solving matrix equations provides the foundation for solving many more problems. In addition to its status as an underlying technology for numerical analysis, we have learned from the least squares application that linear models depend on solving linear equations to find coefficients, and that is the foundation of modern statistical analysis.

Additionally, matrix and vector algebra has applications in science, engineering, and business, making matrix and vector mathematics a common target for numerical analysis researchers. That is why we have so many widely

---

[2]It should be noted that this is an example application and not necessarily the best model for estimating the volume of timber extracted from a tree.

variant methods for solving linear algebra problems. The basic workhorse of these methods, with broadest applicability, will continue to be LU decomposition. Our applications will benefit from the fact LU decomposition prepares the calculation as far as possible, allowing multiple solutions for different outcomes, the $\mathbf{b}$ in $A\mathbf{x} = \mathbf{b}$, a relatively inexpensive computational process. Repeatedly solving this sort of problem is a staple of computer graphics, but the invention of the algorithm dates to the earliest computing pioneers, invented by Alan Turing showing its all around general purpose use.

However, while the LU decomposition is the workhorse of numerical linear algebra, we should not forget the value in specialized solvers. The tridiagonal matrix solver is an especially efficient solution to solving tridiagonal banded matrix problems. Though tridiagonal banded matrices are not especially common, they have a specific application in interpolation for generating cubic splines, as we will see in the next chapter. Other application-specific solvers, such as the iterative methods, only apply to diagonally dominant matrices, but these commonly arise in polynomial optimization problems, making these somewhat constrained matrices appear more frequently than we might expect.

R's `Matrix` library provides a powerful suite of matrix tools, including other decompositions, optimized solvers for several special cases, in addition to classes for storing matrices along with associated metadata. In practice, we would probably extend this library to include new functions if a facility were required that R does not already provide. This allows us to use dependable and reliable functions that have extensive testing as the basis. Beyond `Matrix`, there are a number of specialized packages for linear algebra in R. The `SparsM` package provides support for sparse matrices, for instance. The numerical mathematics task view in CRAN catalogs other packages for solving matrix equations in R.

In addition, R itself relies on linear algebra to implement its core statistical methods. At compile time, and at runtime through packages, there are a number of tools to control how R interacts with the BLAS, the "Basic Linear Algebra Subroutines." The most interesting provide access to higher performance versions of these routines, such as the R packages `RhpcBLASctl` and `bigalgebra`, which gives R a matrix type for large matrices.

Exercises

1. Create a set of functions for the elementary column operations.

2. Assuming A is a 6 by 6 matrix, how many operations are necessary to find $A^\mathsf{T} A$?

3. Solve the following linear system of equations using Gaussian elim-

ination:

$$-4x + 4y = -1$$
$$-2x + 2y - 3z = -3$$
$$3x + 1y - 3z = -3$$

4. R provides %*% and **outer** functions, which implement the dot product and cross product, respectively. Implement your own version of the dot and cross product functions.

5. Find the general form of the solution to upper triangular matrices.

6. Solve the following linear system of equations using LU decomposition:

$$-4x + 4y = -1$$
$$-2x + 2y - 3z = -3$$
$$3x + 1y - 3z = -3$$

7. Write implementations of the forward substitution and backward substitution algorithms for solving triangular matrices.

8. Show the first five iterations of the Jacobi iteration for the following system of equations:

$$3x + 2y - 1z = -3$$
$$-3x - 3y - 3z = 9$$
$$1y - 1z = -1$$

9. Show the first five iterations of the Gauss–Seidel method for the following system of equations:

$$3x + 2y - 1z = -3$$
$$-3x - 3y - 3z = 9$$
$$1y - 1z = -1$$

10. Because the Gauss–Seidel method uses triangular matrices, iterations can be found using forward substitution. Implement that algorithm.

11. The Gauss–Seidel method is the special case of another method called successive over-relaxation. Successive over-relaxation revises equation 3.34, so that,

$$(D + \omega L)\mathbf{x} = \omega b - (\omega U + (\omega - 1)D)\mathbf{x}. \qquad (3.38)$$

Equation 3.38 can be solved for $\mathbf{x}^{(n+1)}$, in the same form as both

Jacobi's method and the Gauss–Seidel method to create an iterative method. With successive over-relaxation, ω is the *relaxation factor* and can be used to increase the rate of convergence. Implement a successive over-relaxation method with a user-settable ω value.

4

Interpolation and Extrapolation

4.1 Polynomial Interpolation

Real-world data presents many problems, one of which is missing data. Data could be missing for a variety of reasons, such as a failed collection, outlier removal, or data was not collected relevant to the point in question. If we have other data, there are a variety of ways to estimate the missing data's value. R users, familiar with statistics, will look toward linear regression, a method we examined in the last chapter. As a staple of statistical analysis, it is powerful and provides well-defined error expectations.

Interpolation, unlike our prior numerical analysis work, is where we get to exercise our intuition and where we might now think "outside of the problem." For instance, given some data points, we can assume those points lie on the same line. Or they may only approximately lie on the same line, one of the fundamental assumptions of linear regression. But the assumption is necessary and some judgment warranted to push the problem along to something we can solve using the analytical toolkit. In some ways, with interpolation, numerical analysis moves from science to art.

This section will explore methods focusing on polynomial interpolation, leading to polynomials of different degrees designed to fit the line. In general, we will look at interpolation and extrapolation in the same way. But they have slightly different definitions. Interpolation is finding the values of missing data in between the known values. Extrapolation extends the model outside of known data to "predict" new values for which no measure could have been taken. The methods for each are similar and distinctions are made only when necessary.

4.1.1 Linear Interpolation

In the first case, we might have only two points. These could be points representing the growth of a child, the amount of sugar in a drink, or the number of homes with televisions in a year. For this kind of real-world data, taking two measurements is cheaper than continuously measuring. So the two measurements are taken at different points in time, or for different sizes of drink, or at different values of whatever it is we are measuring. That measurement

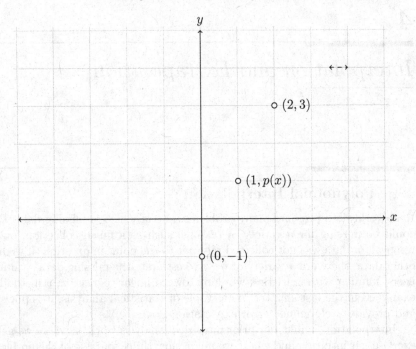

Figure 4.1
Linear interpolation of two points

becomes the x value, which we can determine, and the second measurement, y, is the response variable.

In conventional algebra courses, we solve this problem for two specimen points, (x_1, y_1) and (x_2, y_2) to find an equation of the line in slope-intercept form, $y = mx + b$. Finding the slope, m, requires both ordered pairs, so that,

$$m = \frac{y_2 - y_1}{x_2 - x_1}. \tag{4.1}$$

And the y-intercept, b, is,

$$b = y_2 - mx_2. \tag{4.2}$$

Function 4.1, the `linterp` function, implements these two equations. For two example points, $(2, 3)$ and $(0, -1)$, we would want to find the equation of the line that passes through them. Following from equations 4.1 and 4.2, then

$$m = \frac{-1 - 3}{0 - 2} = 2, \tag{4.3}$$

and,

$$b = y_2 - mx_2 = 3 - 2(3) = -1. \tag{4.4}$$

```
linterp <- function(x1, y1, x2, y2) {
    m <- (y2 - y1) / (x2 - x1)
    b <- y2 - m * x2

    ## Convert into a form suitable for horner()
    return(c(b, m))
}
```

R Function 4.1
Linear interpolation

So the equation of the line defined by these two points is $y = 2x - 1$.

And this is the result returned by the `linterp` function:

```
> (p <- linterp(2, 3, 0, -1))
[1] -1  2
```

The function itself returns two digits. These are the two coefficients representing m and b in the traditional y-intercept form. These numbers are themselves returned as an array suitable for use with the polynomial evaluation functions shown in Section 1.3.2. Here, we can see the result for $x = 1$, using the Horner function for evaluating polynomials in Section 1.3.2:

```
> horner(1, p)
[1] 1
```

And that result appears to agree with the graph in Figure 4.1, so we can feel confident in our result. Of course, the underlying function that generated the two specimen points may not be a line, continuous, or even a function. But that is something we cannot be concerned about at this stage of the analysis.

What we can worry about at this stage is numerical error. There are potential sources for numerical error here in this method if the two specimen points are particularly close together; then the operations $y_2 - y_1$ or $x_2 - x_1$ may result in spuriously large error. This can be a bigger problem in the second subtraction as it is used as a denominator in a fraction.

Aside from potential error, the naïve method is important because it forms the basis of more complex interpolation methods. The first of these is interpolating for higher-order polynomial equations from more than two points.

4.1.2 Higher-Order Polynomial Interpolation

Given two data points, a line, a polynomial of degree one, will always pass exactly through the two points, provided the two values of x are different. If the two values of x are the same, the slope is undefined because it is infinite and the line is vertical.

For three data points, unless those points lie precisely on the same line, no such line is possible. However, a quadratic equation, a polynomial of degree two, will always fit the three data points exactly. And for four data points, a polynomial of degree three will fit. In general, for n data points, a polynomial of degree $n - 1$ is necessary and sufficient to precisely fit the data points. Finding that polynomial is the work of linear algebra. This calculation results in a function, $p(x)$, which is the polynomial approximation of $f(x)$, the source function for the data points.

Given a set of ordered pairs, (x_i, y_i), the interpolating function $p(x)$ must meet the requirement,

$$p(x_i) = y_i, \tag{4.5}$$

for all i. In addition, the interpolating function is a polynomial, therefore it is of the form,

$$p(x) = \beta_n x^n + \beta_{n-1} x^{n-1} + \cdots + \beta_1 x + \beta_0, \tag{4.6}$$

where β_j are the polynomial's coefficients. If we substitute 4.5 into 4.6, we get the following,

$$y_i = \beta_n x_i^n + \beta_{n-1} x_i^{n-1} + \cdots + \beta_1 x_i + \beta_0, \tag{4.7}$$

for each of the i interpolating points. Equation 4.7 can be rewritten in matrix form to,

$$
\begin{bmatrix}
x_1^n & x_1^{n-1} & \cdots & x_1 + 1 \\
x_2^n & x_2^{n-1} & \cdots & x_2 + 1 \\
\vdots & \vdots & & \vdots \\
x_n^n & x_n^{n-1} & \cdots & x_n + 1
\end{bmatrix}
\begin{bmatrix}
\beta_n \\
\beta_{n-1} \\
\vdots \\
\beta_0
\end{bmatrix}
=
\begin{bmatrix}
y_1 \\
y_2 \\
\vdots \\
y_n
\end{bmatrix},
\tag{4.8}
$$

and a matrix equation of the form $X\beta = \mathbf{y}$ is one we can solve. The matrix X is called the Vandermonde matrix and it contains successive columns of x to the nth power. Solving the equation for β returns a vector of values for the coefficients of the polynomial. The Vandermonde matrix and solution are implemented in Function 4.2.

Creating two vectors for the x and y values, respectively, the function `polyinterp` creates the Vandermonde matrix from x and solves for the coefficients. However, if the Vandermonde matrix were created with its columns in the reverse order, the same result can be achieved. Of interest, by reversing the order, the result is usable by the polynomial evaluation functions, like the linear interpolator of the last function. Otherwise, the implementation is straightforward and uses R's internal matrix solving algorithm.

For example, we will assume there are three data points consisting of the order pairs, $(-1, -2)$, $(1, 2)$, and $(0, -1)$. For these points, we will find a quadratic equation, of the form $p(x) = \beta_2 x^2 + \beta_1 x + \beta_0$, such that $p(x_i) = y_i$ for the ordered pairs given. We say that it fits the points and will use $p(x)$ to find the value of $p(-2)$. The data points are shown in Figure 4.2.

Creating vectors for the x values and y values, the function `polyinterp` returns the coefficients of the polynomial which can be evaluated with the **horner** function from Section 1.3.2:

```
polyinterp <- function(x, y) {
    if (length(x) != length(y))
        stop("Length of x and y vectors must be the same")

    n <- length(x) - 1
    vandermonde <- rep(1, length(x))
    for(i in 1:n) {
        xi <- x^i
        vandermonde <- cbind(vandermonde, xi)
    }
    beta <- solve(vandermonde, y)

    names(beta) <- NULL
    return(beta)
}
```

R Function 4.2
Vandermonde polynomial interpolation

```
> x <- c(-1, 1, 0)
> y <- c(-2, 2, -1)
> (p <- polyinterp(x, y))
[1] -1  2  1
```

This result gives the polynomial,

$$p(x) = 1x^2 + 2x + -1, \tag{4.9}$$

and its evaluation at $p(-2)$ returns -1, which does not seem out of place based on the graph given in Figure 4.2:

```
> horner(-2, p)
[1] -1
```

The Vandermonde matrix can also be used to find the line interpolating two points, though it requires far more work than the method of the last section. For the sake of completeness, we can see that the function `polyinterp` provides the same results as the function `linterp` for the same two points:

```
> x <- c(2, 0)
> y <- c(3, -1)
> (p <- polyinterp(x, y))
[1] -1  2
```

In this calculation, the polynomial interpolation via the Vandermonde matrix uses exactly 2 points and returns a polynomial of degree 1.

With polynomial interpolation via the Vandermonde matrix, our worries about the error still include potential floating point calculation error. We must also be aware of potential error coming from the algorithm itself. This is error

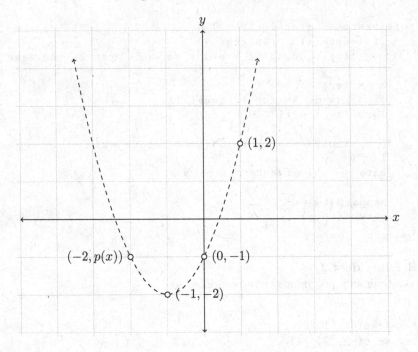

Figure 4.2
Quadratic interpolation of three points

that would exist, even if the calculations underlying the algorithm were carried out with infinite precision.

In the general case, the error for polynomial interpolation is,

$$h(x) = f(x) - p(x) = \frac{(x - x_1)(x - x_2) \cdots (x - x_n)}{n!} f^{(n)}(x), \qquad (4.10)$$

where $p(x)$ is a polynomial of degree $n - 1$ and $\min x_i \leq x \leq \max x_i$ for all i less than n. Of course, $f^{(n)}$ is the nth derivative of the originating function for the interpolated points and that means the error is bounded by some unknown value. So the error in general is not known, unless we already know the underlying function, and that is unfortunate. But if we do know the underlying function, we can calculate the error exactly.

In some cases, we use polynomial interpolation to approximate a known function. For instance, the graph given in Figure 4.3 is based on the tangent function, but it also resembles a cubic function, in some ways. Interpolating among four sample points from this tangent function would provide a cubic polynomial approximation for the tangent function that may ease some other calculation.

One of the biggest drawbacks to pure polynomial interpolation is that the

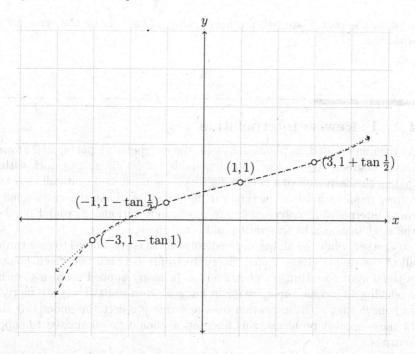

Figure 4.3
Cubic interpolation of $y = \tan((x - 1/4)) + 1$

values tend to diverge from true values at the endpoints, especially when dealing with higher-order polynomials. This leads to wanting a better solution that behaves better at the endpoints. In general, we will seek piecewise approaches that use a number of different polynomials to represent the curve at different points. The simplest is a piecewise linear interpolation we will see next, and later we will look at cubic splines.

Calculating the interpolating polynomial requires passing four observation points directly to the `polyinterp` function.

```
> x <- c(-3, -1, 3, 1)
> y <- c(1 - tan(1), 1 - tan(1/2), 1 + tan(1/2), 1)
> (p <- polyinterp(x, y))
[1]   0.755898927   0.263467854  -0.029050172   0.009683391
```

The coefficients of the polynomial are not round numbers, but as shown in Figure 4.3, the results are close. The dashed line is the "true" value of the function, $f(x) = \tan(x - 1)/4 + 1$. The dotted line represents the interpolating cubic polynomial. Because we know the source function, $f(x)$, in this example, we can calculate the exact error in the interpolating polynomial using equation

4.10. In Chapter 5, we will see an example of calculating the error for our advantage.

4.2 Piecewise Interpolation

As the number of data points increases, the interpolating polynomial grows to higher degrees. Higher-degreed polynomials can be difficult to work with, as finding the derivative of a twelfth degree polynomial, along with all associated terms, requires a lot of mechanical mathematics. That the coefficients for these interpolating polynomials are only rarely rational or round numbers, the mechanics can be frustrating and error-prone.

Further, while the higher-degreed polynomial is guaranteed to pass through all of the points given, it may fluctuate wildly between two given points, a pattern known as Runge's phenomenon. In many applied problems, such as modelling an income curve, we might know *a priori* or have very strong reason to believe that such fluctuation does accurately reflect the underlying data. In these applied problems, this fluctuation should be eliminated to support analysis.

Piecewise interpolation offers an alternative. In piecewise interpolation, we observe that at different points along a curve, a function value may be better approximated using two or more interpolations. We will want to use lower-degreed polynomials for each part of a curve because we are able to efficiently analyze those polynomials to approximately analyze the underlying data.

In the simplest case, we observe that a zero-degreed polynomial is a relatively simple interpolation. This interpolation creates horizontal straight line functions, like $f(x) = c$ for some constant, c, to approximate the function along the way. In the special case where the closest known value is used for the constant interpolation, we call this the nearest neighbor. For instance, given the ordered pairs of $(1, 1)$ and $(3, 3)$, if we wished to interpolate a value for $x = 1.5$, we would observe the closest x value for which we have a value is 1, where the y value is also 1. We would use the value 1 for any value of x closer to 1 than 3. For some data analysis projects, this is sufficient and using either the nearest neighbor, the last value, or the next value, are common in econometric techniques. These slightly different approaches are known collectively as constant interpolation.

But these methods present some problems. The most important of these problems is that the function is not continuous. In the example here, at $x = 2$, there is a jump discontinuity in the value from 1 to 3. Piecewise polynomial interpolation addresses that concern.

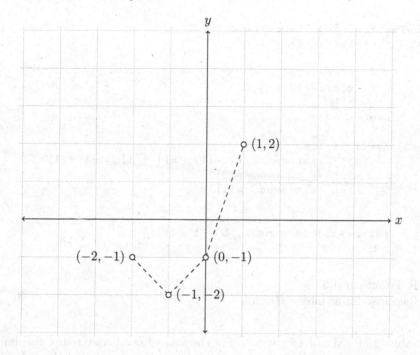

Figure 4.4
Piecewise linear interpolation of four points

4.2.1 Piecewise Linear Interpolation

A piecewise interpolant is of greater value if the resulting function is continuous. A continuous function has smoother transformations from region to region, and that feels more natural. When, for instance, the interpolant represents a real-world function, such as mean income for a country, it is easier to accept a function that transitions between known observation points in an orderly fashion. If the function just jumped from some distance over an infinitesimally small distance, as is the case in nearest neighbor interpolation, then the value added from interpolation is essentially nil.

The initial approach to piecewise interpolation we will look at is piecewise linear interpolation. Piecewise linear interpolation is the linear approach to interpolating between each point. Using Figure 4.2, the graph of a quadratic function, as a reference, we can draw the lines from each point to its neighbor on either side. And at first glance, the interpolating lines are a good first approximation of a quadratic function.

The process for piecewise linear interpolation uses the same process as our linear interpolation model, except it repeats it for each consecutive pair of data points along the x-axis. The function `pwiselinterp` provides an implementation of this process. The function takes two arguments, a vector of x

```
pwiselinterp <- function(x, y) {
    n <- length(x) - 1

    y <- y[order(x)]
    x <- x[order(x)]

    mvec <- bvec <- c()

    for(i in 1:n) {
        p <- linterp(x[i], y[i], x[i + 1], y[i + 1])
        mvec <- c(mvec, p[2])
        bvec <- c(bvec, p[1])
    }

    return(list(m = mvec, b = bvec))
}
```

R Function 4.3
Piecewise linear interpolation

values and a vector of y values. For the sake of good code reuse, this implementation even uses the `linterp` function at its core to find the individual interpolating segments.

The `pwiselinterp` function returns a list of two vectors. One vector, m, is an array of slopes for each component of the piecewise function. The second vector, b, is an array of the corresponding y-intercepts for each component. This implementation does not marry up with the `horner` function evaluator, though it would not be difficult to create an equivalent function evaluator.

```
> x <- c(-2, -1, 0, 1)
> y <- c(-1, -2, -1, 2)
> pwiselinterp(x, y)
$m
[1] -1  1  3

$b
[1] -3 -1 -1
```

The `pwiselinterp` function does not make any assumptions about what happens outside the bounds of the points given. Several strategies can be adopted, depending on the application. One option is to assume that the extreme ends of the piecewise function continue indefinitely. In the case of the quadratic function, that would even be reasonable for a short distance. Another option is to flatten out the "curve" and assume that y is equal to the end measurements for any otherwise undefined function value. Of course, it is also possible to leave the values undefined.

R provides a function, `approx`, that provides piecewise linear interpolation, by default. This function works by accepting a list of points to interpolate through, and a list of points for approximation values. So the `approx` function, instead of returning the parameters of a piecewise function, instead returns y values associated with the points in question. In addition, the `approxfun` returns an executable function that provides the piecewise interpolation for any arbitrary x value.

```
> f <- approxfun(x, y)
> f(0)
[1] -1
> f(0.5)
[1] 0.5
```

Like `approx`, the `approxfun` function provides linear piecewise interpolation by default. Both accept a parameter, `method`, which can return a constant piecewise function with the value "constant."

Typically, in calculus, a function that is analyzed would need to be both continuous and differentiable. When a function is differentiable, it is not just continuous. It also has a defined derivative at each point. In practice, this means there are no sharp corners. As we can see from Figure 4.4, there's a sharp corner at every point when the two interpolating lines meet. But the function is still continuous. Splines give us greater power to create interpolants that are not just continuous but also differentiable.

4.2.2 Cubic Spline Interpolation

If using a first-degree polynomial, a line, over multiple intervals is an improvement over a single interpolating line, and if using a higher-degree polynomial were also an improvement over a single interpolating line, then we might infer that using higher-degree polynomials over multiple intervals would also be an improvement. In some cases, this would be correct, but we would still face sharp corners where each of the interpolating curves joined. These sharp corners prevent differentiation and, more practically, cannot be used to model some real-world functions, such as a roller coaster span.

The cubic spline solves this problem. Any piecewise interpolation may be referred to as a spline. But without qualification, we are normally talking about the cubic spline. The cubic spline, while piecewise, provides a differentiable and smooth curve, despite the joins. That also makes the spline integrable. As each individual section is represented by a cubic curve, a polynomial of degree 3, then each individual section can also be analyzed as a cubic curve. Assuming there are n data points to interpolate, we will define S_i as the cubic polynomial function representing the curve over the domain $[x_i, x_{i+1}]$. Then for n data points, there are $n - 1$ interpolating cubic polynomials.

Initially, the series of polynomials can be generalized so that,

$$S_i = d_i(x - x_i)^3 + c_i(x - x_i)^2 + b_i(x - x_i) + a_i. \tag{4.11}$$

This leads to 4 unknowns, a_i, b_i, c_i, and d_i for each equation. Therefore, there are $4(n-1) = 4n - 4$ unknown values. Because we want the spline to be continuous and differentiable, there is a set of equations that define the cubic spline:

$$S_i(x_i) = y_i, \qquad\qquad i = 1, \ldots, n-1 \qquad (4.12)$$
$$S_i(x_{i+1}) = y_{i+1}, \qquad\qquad i = 1, \ldots, n-1 \qquad (4.13)$$
$$S_i'(x_i) = S_{i+1}'(x_i), \qquad\qquad i = 1, \ldots, n-2 \qquad (4.14)$$
$$S_i''(x_i) = S_{i+1}''(x_i), \qquad\qquad i = 1, \ldots, n-2 \qquad (4.15)$$

Equations 4.12 and 4.13 are clear. These requirements ensure that if we evaluate our spline at one of the internal nodes, which we sometimes call knots, our result is the predetermined answer. That is, the spline evaluated at x_i, for some i, is y_i, and each component of the spline joins up neatly. Equation 4.14 ensures that we have a continuous first derivative at each internal node. This prevents there from being a corner at the node. Equation 4.15 ensures a continuous second derivative, which is primarily advantageous because it means the first derivative is itself differentiable.

These conditions lead to $4n - 6$ conditions we must meet. If our $4n - 4$ unknowns were solved as a matrix, and eventually they will be, the matrix would be underdetermined. We can resolve this underdetermination by including two addition conditions. With cubic splines, the normal behavior is to specify the end conditions on either end to achieve two additional conditions. For this example, the two conditions we will add are $S''(x_1) = 0$ and $S''(x_n) = 0$. These two conditions ensure that at the endpoints, the first derivative is linear and therefore the spline function continues off in the direction it was already going. We call this the "natural spline."

Initially, we can see that each S_i cubic polynomial is right-shifted by x_i units or left-shifted if x_i is negative. At x_i, the value of the function is a_i, which means that $a_i = y_i$ for each value of i. Resolving the rest of the coefficients is more complex, but now it is $3n$ unknowns with $3n$ conditions. The full derivation is available from multiple sources, but for an outline, if from equation 4.14 we get $S_i'(x) = S_{i+1}'$, then $S_{i+1}' - S_i'(x) = 0$, and we can substitute 4.11 into both components, solving for d_i in terms of x_i, y_i, and c_i The same process can be replicated with equation 4.15 and b_i. The result is a tridiagonal matrix, just like those from Section 3.2.2, that can be solved to find the coefficients. There is a matrix, A, such that,

$$AC = V \qquad (4.16)$$

where A is a tridiagonal matrix. In this tridiagonal matrix, $(u_1, u_2, \ldots, u_{n-1})$ is a vector $\mathbf{U}$, and the vectors $\mathbf{M}$, $\mathbf{L}$, and $\mathbf{V}$ are similarly defined. These matrices are such that,

$$u_i = l_i = x_{i+1} - x_i, \qquad (4.17)$$

except that $u_0 = l_n = 0$. Further, the main diagonal, $\mathbf{D}$, is defined by,

$$m_i = 2(x_{i+1} - x_i + x_i - x_{i-1}), \qquad (4.18)$$

except that $d_0 = d_n = 1$. Finally, $\mathbf{V}$ is defined by

$$v_i = 3\left(\frac{y_{i+1} - y_i}{x_{i+1} - x_i} - \frac{y_i - y_{i-1}}{x_i - x_{i-1}}\right), \tag{4.19}$$

except that $v_0 = v_n = 0$. Therefore,

$$\begin{bmatrix} m_1 & u_1 & 0 & 0 & 0 \\ l_1 & m_2 & u_2 & 0 & 0 \\ 0 & l_2 & \ddots & \ddots & 0 \\ 0 & 0 & \ddots & \ddots & u_{n-1} \\ 0 & 0 & 0 & l_{n-1} & m_n \end{bmatrix} \mathbf{C} = \begin{bmatrix} v_1 \\ v_2 \\ \vdots \\ \vdots \\ v_n \end{bmatrix}, \tag{4.20}$$

Solving for $\mathbf{C}$ yields the c vector. From here, the b vector is,

$$b_i = \frac{y_{i+1} - y_i}{x_{i+1} - x_i} - \frac{x_{i+1} - x_i}{3}(2c_i + c_{i+1}), \tag{4.21}$$

and

$$d_i = \frac{c_{i+1} - c_i}{3(x_{i+1} - x_i)}. \tag{4.22}$$

We can solve for $\mathbf{C}$ using the tridiagonal matrix and the vectors $\mathbf{B}$ and $\mathbf{D}$ can be found using $\mathbf{C}$.

The cubic spline function, available as `cubicspline` in Function 4.4 implements this algorithm, using the `tridiagmatrix` function to manage the bulk of the execution, ensuring an efficient implementation. The `cubicspline` function itself returns a list of four vectors, named a, b, c, and d, corresponding to the same variables names in equation 4.11. Using the four vectors, each cubic polynomial can be assembled with coefficients.

Using the same example points as the linear piecewise interpolation example, we can see the cubic spline in action:

```
> x <- c(-2, -1, 0, 1)
> y <- c(-1, -2, -1, 2)
> cubicspline(x, y)
$a
[1] -1 -2 -1

$b
[1] -1.4 -0.2  2.2

$c
[1] 0.0 1.2 1.2

$d
[1]  0.4  0.0 -0.4
```

```
cubicspline <- function(x, y) {
    n <- length(x)
    dvec <- bvec <- avec <- rep(0, n - 1)
    vec <- rep(0, n)
    deltax <- deltay <- rep(0, n - 1)

    ## Find delta values and the A-vector
    for(i in 1:(n - 1)) {
        avec[i] <- y[i]
        deltax[i] = x[i + 1] - x[i]
        deltay[i] = y[i + 1] - y[i]
    }

    ## Assemble a tridiagonal matrix of coefficients
    Au <- c(0, deltax[2:(n-1)])
    Ad <- c(1, 2 * (deltax[1:(n-2)] + deltax[2:(n-1)]), 1)
    Al <- c(deltax[1:(n-2)], 0)

    vec[0] <- vec[n] <- 0
    for(i in 2:(n - 1))
        vec[i] <- 3 * (deltay[i] / deltax[i] -
                            deltay[i-1] / deltax[i-1])

    cvec <- tridiagmatrix(Al, Ad, Au, vec)

    ## Compute B- and D-vectors from the C-vector
    for(i in 1:(n-1)) {
        bvec[i] <- (deltay[i] / deltax[i]) -
            (deltax[i] / 3) * (2 * cvec[i] + cvec[i + 1])
        dvec[i] <- (cvec[i+1] - cvec[i]) / (3 * deltax[i])
    }

    return(list(a = avec, b = bvec,
                c = cvec[1:(n - 1)], d = dvec))
}
```

R Function 4.4
The natural cubic spline

For instance, over the domain from $[0,1]$, the cubic spline is $-0.4x^3 + 1.2x^2 + 2.2x - 1$, a relatively mundane polynomial. In addition, it can be differentiated multiple times using nothing but the power and multiplication rules, since no trigonometric or exotic functions appear in the equation. Therein lies the strength of the cubic spline in providing a model for data. This is illustrated in Figure 4.5.

In Figure 4.6, we see a cubic spline interpolation of five points that are

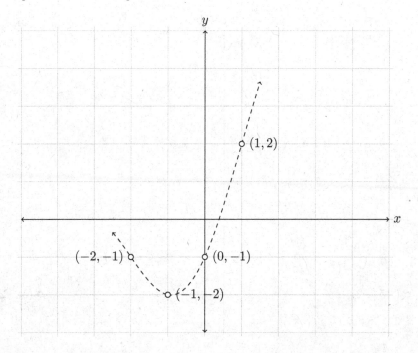

Figure 4.5
Cubic spline interpolation of four points

symmetric about the y-axis. As we might expect, if the points are symmetric, so is the spline interpolation. However, this graph also shows the spline interpolation is not necessarily a straight path. In the first section of the spline, the interpolating polynomial bends downward before coming up again to meet the right-side knot. The last section shows the same behavior in mirror image. As a cubic polynomial, the curve will bend at least twice and it is possible one of those bends will be in the interpolated region. Accordingly, we cannot rely on the spline behaving strictly monotonically between knots, as the spline in Figure 4.5 does.

One design choice, reinforced in this discussion, is that the `cubicspline` function can only generate natural splines. There are other options for the endpoints of the cubic spline, including clamping, which flattens out the ends to level lines or some generate periodic functions. These are created by changing the end conditions, which our implementation sets to create a straight line in the direction of the line at the boundary point.

While correct, this implementation of the cubic spline function is unwieldy, as the result describes a series of parameters that describe the functions comprising the spline. Within R, there is a built-in function called `splinefun` in the `stats` package that can create a function that represents the spline. The

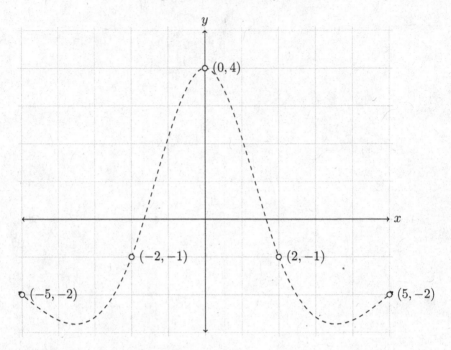

Figure 4.6
Cubic spline interpolation of five points

results of this operation is a new function that can be executed and provides the spline function, alleviating the concern about interpreting and applying the coefficients. Using the existing x and y values from Figure 4.6,

```
> spline.ex <- splinefun(x, y, method = "natural")
> spline.ex(x)
[1] -2 -1  4 -1 -2
```

And the new `spine.ex` function returns the exact values of y expected at the known x points.

4.2.3 Bézier Curves

Bézier curves provide a completely different alternative to the polynomial spline curves we just considered. Like other interpolating functions, Bézier curves are polynomials and a Bézier curve defined by n data points will be a polynomial function of $n - 1$ degree, like the results of the Vandermonde matrix approach to polynomial interpolation. But rather than passing through the n points, a Bézier curve passes through the first and last point and the intermediate points provide guides to control the shape of the curve, without the requirement that the curve pass through the guide points.

Bézier curves have a number of applications in computer graphics and imaging. Bézier tools are commonly available in vector graphics programs, logo design tools, animation programs, and font design toolkits. These applications sometimes refer to the Bézier curve as a path. These applications, and others, benefit from smooth curves and the relatively fast approach to calculating values. In addition, a high-definition curve can be defined by a handful of relevant data points, allowing a designer to create a complete curve with a large amount of embedded data with only a few numbers needed for reconstruction.

Bézier curves are parametric. Instead of the value y being some function of x, both y and x are functions of some third variable, t. Mathematically, each point is defined as $(x(t), y(t))$. This will provide a number of interesting features. For instance, because y is not necessarily a function of x, the curve may loop backwards or even cross itself, provided that both $x(t)$ and $y(t)$ are each valid functions of t. Because each is a polynomial function of t, the x and y positions can be calculated quickly using standard polynomial functions. Further, the $x(t)$ and $y(t)$ functions can be interpreted as their own functions, with derivatives and so forth found for each.

We will start by assuming there are n points $P = \{p_1, \ldots, p_n\}$ where $p_1 = (x_1, y_1)$, $p_2 = (x_2, y_2)$, and so forth to $p_n = (x_n, y_n)$. Bézier curves are defined so that the slope at p_1 is tangent to the line between p_1 and p_2 and the slope at p_n is tangent to the line at p_{n-1} and p_n. However, the slope at the endpoints is not defined; additional points to the setting allow the slopes to be set. The simplest nontrivial Bézier curve is the quadratic Bézier curve, for a given set of points, (x_1, y_1), (x_2, y_2), and (x_3, y_3); then,

$$x(t) = (1-t)^2 x_1 + 2t(1-t)x_2 + t^2 x_3 \qquad (4.23)$$
$$y(t) = (1-t)^2 y_1 + 2t(1-t)y_2 + t^2 y_3 \qquad (4.24)$$

where $t \in [0, 1]$. Therefore, $(x(0), y(0)) = (x_1, y_1)$ and $(x(1), y(1)) = (x_3, y_3)$. These formulae are implemented in Function 4.5, `qbezier`.

The motivation for developing Bézier curves is different from our other piecewise interpolation routines. In those routines, we are attempting to mimic the behavior of a dataset, or a curve, in an attempt to analyze the dataset or curve by proxy. With Bézier curve, our principal motivation is to draw pictures of the curve. Accordingly, this implementation obliges and instead of returning the parameters of an interpolating curve, provides point estimates for positions along the curve.

The function accepts three parameters. The first is a vector of x values and the second is a vector of corresponding y values. Obviously, these must both have exactly three entries. The last value is a collection of t values, for which the curve's points should be calculated. The `qbezier` function returns a list of two vectors, one with new x values and one with new y values. Ordered pairs can then be generated from the list.

```
qbezier <- function(x, y, t) {
    if(length(x) != 3 || length(y) != 3)
        stop("x and y must contain exactly 3 values")

    newx <- (1-t)^2 * x[1] + 2 * (1-t) * t * x[2] +
        t^2 * x[3]
    newy <- (1-t)^2 * y[1] + 2 * (1-t) * t * y[2] +
        t^2 * y[3]

    return(list(x = newx, y = newy))
}
```

R Function 4.5
Quadratic Bézier curves

```
> x <- c(1, 2, 3)
> y <- c(2, 3, 5)
> qbezier(x, y, seq(0, 1, 1/5))
$x
[1] 1.0 1.4 1.8 2.2 2.6 3.0

$y
[1] 2.00 2.44 2.96 3.56 4.24 5.00
```

The image drawn from this sample data can be seen in Figure 4.7, though the image could not be drawn from the five sample values of x and y produced. Since the purpose of working with Bézier curves is to produce curves that are generally pleasing to the eye, a sufficient number of data points should be produced to render a picture that looks like a curve without jagged edges. This requires experimentation and is subject to the implementor's discretion.

A cubic Bézier curve is only slightly more complicated than the quadratic Bézier curve. For a given set of points, (x_1, y_1), (x_2, y_2), (x_3, y_3), and (x_4, y_4), then,

$$x(t) = (1 - t)^3 x_1 + 3t(1 - t)^2 x_2 + 3(1 - t)t^2 x_3 + t^3 x_4 \qquad (4.25)$$

$$y(t) = (1 - t)^3 y_1 + 3t(1 - t)^2 y_2 + 3(1 - t)t^2 y_3 + t^3 y_4 \qquad (4.26)$$

where $t \in [0, 1]$. Therefore, $(x(0), y(0)) = (x_1, y_1)$ and $(x(1), y(1)) = (x_4, y_4)$. These formula are implemented in Function 4.6, cbezier.

Like the qbezier function, the cbezier function expects three vectors for arguments, with x and y each providing four elements for four corresponding points.

```
> x <- c(-1, 1, 0, -2)
> y <- c(-2, 2, -1, -1)
> cbezier(x, y, seq(0, 1, 1/5))
```

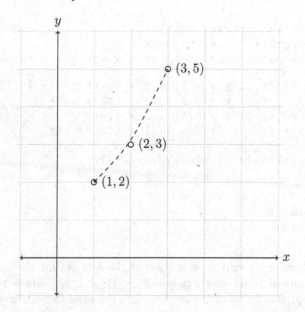

Figure 4.7
Quadratic Bézier interpolation of three points

```
$x
[1] -1.000 -0.144  0.088 -0.208 -0.936 -2.000

$y
[1] -2.00 -0.36  0.08 -0.20 -0.72 -1.00
```

The image drawn from the sample points can be seen in Figure 4.8. We can also see how the property of the curve passing over itself can be achieved using the same data, but without enforcing the order requirement necessary for the cubic spline function.

```
> x <- c(-1, 0, 1, -2)
> y <- c(-2, -1, 2, -1)
> cbezier(x, y, seq(0, 1, 1/5))
$x
[1] -1.000 -0.432 -0.056 -0.064 -0.648 -2.000

$y
[1] -2.000 -1.224 -0.352  0.232  0.144 -1.000
```

The results of this new Bézier curve are shown in Figure 4.9, which is identical to Figure 4.8 except the control points are swapped.

In the general form, a Bézier curve can be calculated for any degree poly-

```
cbezier <- function(x, y, t) {
    if(length(x) != 4 || length(y) != 4)
        stop("x and y must contain exactly 4 values")

    newx <- (1-t)^3 * x[1] + 3 * (1-t)^2 * t * x[2] +
        3 * (1-t) * t^2 * x[3] + t^3 * x[4]
    newy <- (1-t)^3 * y[1] + 3 * (1-t)^2 * t * y[2] +
        3 * (1-t) * t^2 * y[3] + t^3 * y[4]

    return(list(x = newx, y = newy))
}
```

R Function 4.6
Cubic Bézier curves

nomial. There is a recursive definition of the formula for the curve. The formula provides that for given points $P = \{p_1, \ldots, p_n\}$ where $p_1 = (x_1, y_1)$, $p_2 = (x_2, y_2)$, and so forth to $p_n = (x_n, y_n)$, then the Bézier curve for $t \in [0, 1]$ is,

$$B(t, P) = \begin{cases} p_1 & \text{if } n = 1 \\ (1 - t)B(t, P_\alpha) + tB(t, P_\beta) & \text{otherwise} \end{cases}, \quad (4.27)$$

where $P_\alpha = \{p_1, \ldots, p_{n-1}\}$ and $P_\beta = \{p_2, \ldots, p_n\}$. The point may be treated as a vector or the x and y values processed separately. The method also extends to higher dimensional Bézier curves. This recursive definition can scale up to any degree polynomial, $n - 1$, given a sufficient number of data points. However, as the derived coefficients increase in size, the problem becomes numerically unstable for sufficiently large values of n. One approach to address this instability is to create independent quadratic or cubic Bézier curves piecewise over a complete set of points, p.

When considering a Bézier curve, the functional form shown in equations for the quadratic and cubic Bézier can be misleading. Normally, we see functions as some operation on x to produce a value. Here, $x_1, \ldots, x_n$ are constants to be filled in, based on the available data, and the operation is on t. The mathematics does not change, but the expectation does and being mindful of that expectation is important to managing the application.

Like the spline function, R has a built-in function for producing Bézier curves. The `grid.bezier` function, part of the `grid` package of graphics tools, can produce Bézier curves for any number of data points, given by x and y values. However, like this implementation, the primary purpose of the `grid.bezier` function is to draw the graphic, not create analyzable functions. Accordingly, the result is an image and not a function.

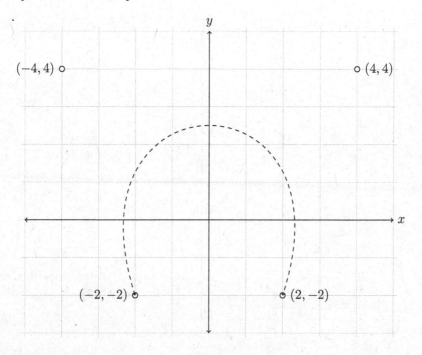

Figure 4.8
Cubic Bézier interpolation of four points

4.3 Multidimensional Interpolation

Most datasets are not strictly univariate nor functions of single variables. Often, we need to interpolate across multiple dimensions. This is useful in data science and statistics, when we need to interpolate values for data process. It is also useful to interpolate in multiple dimensions for computer graphics. We will begin this look by extending our elementary univariate interpolation methods to multiple dimensions. Then we will take a step backwards and look at nearest neighbor interpolation.

4.3.1 Bilinear Interpolation

Extending the basic idea of linear interpolation to two dimensions is bilinear interpolation. Bilinear interpolation is suitable when we have a function $z = f(x, y)$, and we have four known values of z with associated x and y values. Then, we can interpolate new values of z for intermediating values of x and y. There are some limitations on this method. First, we expect that the four known values form a rectangle in two-dimensional space. That is, the points are

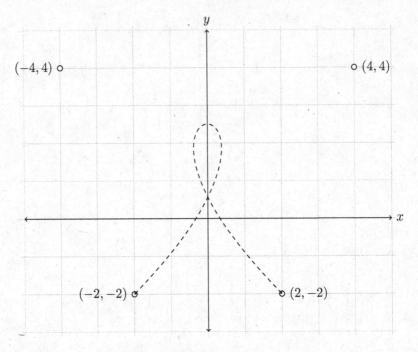

Figure 4.9
Cubic Bézier interpolation of four points, with crossover

defined as (x_1, y_1), (x_2, y_1), (x_1, y_2), and (x_2, y_2). Second, we expect values of z to be defined at each of those points. This allows us to take a straightforward approach to interpolation in two dimensions.

The basic approach is that we initially interpolate along the x-axis between x_1 and x_2. We have to perform this step twice, once along y_1 and once along y_2. Accordingly, we start by defining intermediating values of z, so that,

$$z_1' = \frac{(x_2 - x)z_{1,1} + (x - x_1)z_{1,2}}{x_2 - x_1} \tag{4.28}$$

$$z_2' = \frac{(x_2 - x)z_{2,1} + (x - x_1)z_{2,2}}{x_2 - x_1} \tag{4.29}$$

where (x, y) is the target interpolation point. Once these intermediate values of z are found, the next step is to interpolate between them along the y-axis. Therefore,

$$z = (y_2 - y)z_1' + (y - y_1)z_2' \tag{4.30}$$

There's nothing unusual happening in these formulae except that the values of z_1' and z_2' are presumed to be correct and a valid starting point for the second round of interpolation. Otherwise, the principle is the same as that shown in `linterp`, and is linear in two dimensions.

```
bilinear <- function(x, y, z, newx, newy) {

    ## Find intermediate values along the x-axis, first
    z1 <- (x[2] - newx) * z[1,1] + (newx - x[1]) * z[1,2]
    z1 <- z1 / (x[2] - x[1])
    z2 <- (x[2] - newx) * z[2,1] + (newx - x[1]) * z[2,2]
    z2 <- z2 / (x[2] - x[1])

    ## Then interpolate along the y-axis
    z <- (y[2] - newy) * z1 + (newy - y[1]) * z2
    z <- z / (y[2] - y[1])

    return(z)
}
```

R Function 4.7
Bilinear interpolation

The `bilinear` function, shown in Function 4.7 is the implementation of this procedure. This implementation sets aside several performance enhancements in favor of the simplified structure that mirrors equations 4.28, 4.29, and 4.30. In addition, unlike the implementation of `linterp`, the `bilinear` function does not return the terms of a generalized function. Instead, it accepts two arrays of x and y values where the function is to be evaluated. The `bilinear` interpolates the points at each of the ordered pair and returns a vector of z values.

Imagine a set of four points with the following properties: $f(2,4) = 28$, $f(4,4) = 34$, $f(2,7) = 31$, and $f(4,7) = 35$. These values can represent many different types of physical phenomena, such as discrete temperature measurements at the four points, lighting intensity, or the height of some structure. For now, we will assume these are temperature measurements. The points and measurements are shown in Figure 4.10.

It may be desirable to interpolate values of the temperature among the four points to create a temperature gradient. In our case, we create the x and y vectors, along with a matrix, z, that contains the z values for the four ordered pairs. It is, strictly, not necessary to store the z values in a matrix, but it does create a kind of conceptual simplicity. Finally, two new vectors of interpolating ordered pairs are created and the results produced. We can see this process here.

```
> x <- c(4, 7)
> y <- c(2, 4)
> z <- matrix(c(28, 31, 34, 35), nrow = 2)
> newx <- c(2.5, 3, 3.5)
```

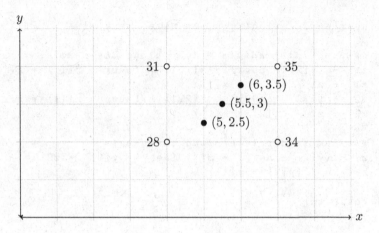

Figure 4.10
Bilinear interpolation example

```
> newy <- c(5, 5.5, 6)
> bilinear(x, y, z, newx, newy)
[1] 31.00000 32.41667 33.66667
```

The straightforward approach has some drawbacks. The most substantial is that there is no continuity across regions of interpolation. That is, provided there are two sets of bounding boxes that share two common points, so they share a vertex, there would be a sharp edge along the vertex. This sharp edge is a two-dimensional analog of the corners found in piecewise linear interpolation, as this is a sort of regionwise bilinear interpolation. The other substantial drawback is, like with linear interpolation, there is no sense of what is happening within the interpolated region. There is only the assumption that an interpolated region moves from bounding point to bounding point in a smoothly monotonic manner, an assumption that may not fit the real world.

The largest drawback to bilinear interpolation is that the method necessarily works on two-dimensional gridded data, like a picture, or more abstractly, like a spreadsheet. In other words, the method works on data that exists in a rectilinear grid and contains a single observation at each point on the grid, where the observation is a third dimension on the grid. However, the method is very fast and it is quite extensible. It can be extended to a four-dimensional implementation, where $z = f(w, x, y)$ and we wish to interpolate across three dimensions. This method, called *trilinear* interpolation, works similarly to bilinear interpolation, with an extra set of steps added to account for the additional dimension.

Another substantial drawback is the interpolated data must be rectilinear. The method assumes there are observed data points across the grid, regardless of the dimension count. Missing values could be filled in using other methods, but the bilinear method is not suitable for sparse data, or data that is missing a substantial number of observations. We will see in the next subsection a method, nearest neighbor, that is better for very sparse data across highly dimensional datasets.

Despite these limitations, bilinear interpolation has important applications in the field of computer graphics, where the constraints are fairly easily met. In computer vision, bilinear interpolation can be used to resample an image and fill in details that were not scanned. Bilinear interpolation can also be used to resize an image, like when we use Photoshop, or Apple Preview to change the dimensions of an image. This will be shown in Section 4.4.2. Further, in three-dimensional graphic rendering, bilinear interpolation can be used to improve the realism of a graphic that is applied to a flat surface, a process known as texture mapping.

4.3.2 Nearest Neighbor Interpolation

While bilinear interpolation, and its extensions to more dimensions, are excellent methods for fast interpolation on gridded data, not all data is gridded, with nearly complete observations at each possible data point. For instance, a dataset of observations from a sociological study will have many dimensions such as age, location, education level, income level, and so forth. There will not be a single observation for each possible combination across all of the dimensions, as is required for perfectly gridded data. There may, in fact, be multiple observations for some combinations across dimensions. Accordingly, we also need methods for interpolating missing observations across sparse datasets.

Nearest neighbor may be the simplest possible option for sparse data interpolation. With nearest neighbor, we start with a point p, that we wish to interpolate a value for. Then we consider nearby points for which we already have valid observations and identify the single closest point, p_1, usually through Euclidean distance. Then we substitute the value of p_1 for p.

In one dimension, this naive process produces the piecewise constant interpolation, which is appropriate in some applications. Consider the example of $f(x) = x$, where we have values for $x = 1, 2, 3,$ and 4. If we interpolate along the domain of $x \in [1, 4]$ using nearest neighbor, then we will end up with a piecewise series of constant lines that resemble a staircase, as shown in Figure 4.11.

While this approach is sometimes used in modeling linear functions, it leaves us looking for a more robust solution. The biggest problem with nearest neighbor in the staircase obliterates the clear upward trend in the data. Of course, we have more advanced tools for modeling these functions, such as Vandermonde polynomials and piecewise interpolators. But in more than one dimension, the nearest neighbor provides a more resilient method for a lower computational cost than other solutions.

Consider the example of two-dimensional nearest neighbor applied to the same data as Figure 4.10. Shown in Figure 4.12, the grid is divided into four

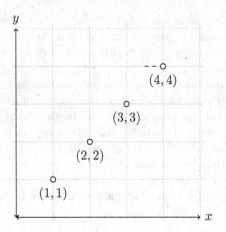

Figure 4.11
Nearest neighbor interpolation in one dimension

regions, labeled A, B, C, and D, where the nearest neighbor is the labeled point within each region. For any point within a given region, the value will be the same as the labeled point. In a sense, this is a step backwards from bilinear interpolation. However, the algorithm is fast if a bit coarse. As additional dimensions are added, nearest neighbor gains speed advantages.

In data science, nearest neighbor is commonly used to address a missing data problem. Missing value imputation is critical in data science as many predictive algorithms, such as logit or random forest, cannot process a missing value. And if there are relatively few missing values in a given column, then imputing a value from existing data is one solution. Because many data science problems use dozens or even hundreds of columns of input data, calculating the nearest neighbor is relatively quick and feels like a more natural approach than, for instance, hot deck imputation.

The function **nn**, shown in Function 4.8 implements a simple n-dimensional nearest neighbor search. It accepts three parameters. The first, p, is a matrix of n columns and m observations. The second, y, is a vector of values for some additional column of data that we wish to interpolate. The third value, q, is a vector of n entries corresponding to the n columns of the source dataset. The function works by replicating the q vector into a new matrix with the same number of "observations" as p. Then the elementwise difference is taken, squared, and the rowsums found. Finally, the square root of the rowsum is taken, though it is completely redundant. The minimum of the rowsums and the minimum of the square root of the rowsums will be identical. Finally, the function returns the corresponding element of y. We can see this in action on a set of random integers.

```
> (p <- matrix(floor(runif(20, 0, 9)), 4))
```

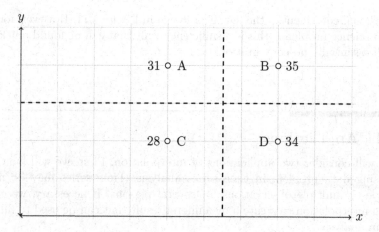

Figure 4.12
Two-dimensional nearest neighbor example

```
nn <- function(p, y, q) {
    if(ncol(p) != ncol(q))
        stop("p and q must have same number of columns")

    ## Repeat the rows of q to simplfy the   calculation
    qprime <- t(matrix(rep(q, nrow(p)), ncol(p)))
    d <- sqrt(rowSums((p - qprime)^2))
    return(y[which.min(d)])
}
```

R Function 4.8
Nearest neighbor interpolation

```
      [,1] [,2] [,3] [,4] [,5]
[1,]    5    6    3    3    4
[2,]    4    7    2    2    7
[3,]    2    5    6    3    0
[4,]    1    0    5    4    8
> (y <- floor(runif(4, 0, 9)))
[1] 6 2 4 5
> (q <- matrix(floor(runif(5, 0, 9)), 1))
      [,1] [,2] [,3] [,4] [,5]
[1,]    6    7    5    6    7
> nn(p, y, q)
[1] 6
```

The nearest neighbor is a constant interpolator and will return a multidimensional equivalent of the flat lines shown in Figure 4.11. However, for many data science problems, this is a sufficient result that can be found quickly and with essentially no computational error.

4.4 Applications

We will consider two applications of interpolation. First, we will look at the missing data problem in time series analysis. Time series may be missing single or multiple observations for several reasons. If necessary, we can use data interpolation to create a meaningful result that can be used to fill a hole in an analysis.

In the second case, we will look at how we can use interpolation to resize images for display. Some of these methods are plain and approachable, while keeping a high speed. Others produce better results by spending more time on the image and applying more complicated methods. The last subsection in this chapter will explore those different methods and their use.

4.4.1 Time Series Interpolation

Interpolation holds an important place specifically in time series analysis. Time series analysis focuses on single-dimensional data streams that are governed by a time index. Time series arise in physics, social science, and other fields. Time series analysis has been critical in the development of modern economics and finance. Time series can represent unemployment on a monthly basis or could represent trading data, where prices pair with times for how much a stock costs. Typically, a time series uses a regular polling interval, such as once a week, once a month, or once a quarter. However, some time series, such as the trading data, may be irregularly spaced. Working with time series sometimes forces the analyst to interpolate a reasonable value to fit into an analysis.

Within the limited domain of time series, there are a number of methods we can use to interpolate values. For the purpose of this explanation, we are going to assume we are working with a time series that is regular and is missing one or more observations. This simplifies discussion, and the underlying methods can be extended to fill in estimates for unobserved dates. This includes period conversation, such as converting from a quarterly time series, with four observations per year, to a monthly time series, where there are 12 per year.

We will also assume we are using one of the internal representations of time series. R includes a package called zoo which can represent time series. In addition, a package called xts is also available for download. The xts package extends zoo, and nearly everything said about zoo also applies to xts. Therefore, we will use zoo as the reference package.

We start by loading the zoo library and creating a time series object. This data is the daily high temperature, in Celsius, at the Maryland Science Center in Baltimore, Maryland. The series itself covers June of 2015. Several daily observations have been removed and set to NA to facilitate understanding this time series and interpolation.

```
> library(zoo)
> (msc <- read.zoo("msc-temps.csv", "%d-%b-%y", sep = ","))
2015-06-01 2015-06-02 2015-06-03 2015-06-04 2015-06-05
      30.0       19.4       17.8         NA       22.8
2015-06-06 2015-06-07 2015-06-08 2015-06-09 2015-06-10
      26.7       25.0       31.1       30.0       29.4
2015-06-11 2015-06-12 2015-06-13 2015-06-14 2015-06-15
      35.0       35.0         NA       31.7       32.2
2015-06-16 2015-06-17 2015-06-18 2015-06-19 2015-06-20
      33.9       29.4       30.0       30.0       30.6
2015-06-21 2015-06-22 2015-06-23 2015-06-24 2015-06-25
        NA         NA       36.1       29.4       28.3
2015-06-26 2015-06-27 2015-06-28 2015-06-29 2015-06-30
        NA         NA         NA       28.3       30.6
```

We use **read.zoo** to read the data from a CSV file. The first option is the file name and the second specifies the date format. This date format presents a date in the form of "15-Jun-2015."

When interpolating a time series, it is convention to think of it as a sequence of events, since the series normally consists of a sequence of discrete measurements that are ordered. We can think of observations as next or previous, later or earlier. That thinking leads directly to the first interpolation approach, the zero-order hold. The zero-order hold assumes that the last valid measurement we saw continues until we receive a new valid measurement. Therefore, all intervening observations that should be filled in should be filled in with the first prior observation found. The zoo package implements this method directly and calls it **na.locf**, for "last observation carried forward."

```
> (msc.locf <- na.locf(msc))
2015-06-01 2015-06-02 2015-06-03 2015-06-04 2015-06-05
      30.0       19.4       17.8       17.8       22.8
2015-06-06 2015-06-07 2015-06-08 2015-06-09 2015-06-10
      26.7       25.0       31.1       30.0       29.4
2015-06-11 2015-06-12 2015-06-13 2015-06-14 2015-06-15
      35.0       35.0       35.0       31.7       32.2
2015-06-16 2015-06-17 2015-06-18 2015-06-19 2015-06-20
      33.9       29.4       30.0       30.0       30.6
2015-06-21 2015-06-22 2015-06-23 2015-06-24 2015-06-25
      30.6       30.6       36.1       29.4       28.3
```

2015-06-26	2015-06-27	2015-06-28	2015-06-29	2015-06-30
28.3	28.3	28.3	28.3	30.6

Of course, this method has some drawbacks. The most important drawback is the sudden jump when a new observation is made. There is no suggestion of a smooth transition between two observations. While this has all of the attendant problems previously detailed with jump discontinuity in interpolation, the method is commonly used. That may be acceptable when we assume a continuing observation which may be unwarranted in practice. For instance, when we think about how much it costs to fill a car with gas, we tend to think of the last price we saw, though it has likely changed since then. Or if we are analyzing some socioeconomic data, we fix the value of unemployment at the observed value before the data was collected. So many time series make few large jumps in a short time, that carrying an observation through is sufficiently close to be reasonable. The `na.locf` function also includes a version that works in reverse, a "next observation carried backwards" version.

The `zoo` package includes several other interpolators. One, `na.approx` creates linear approximations. Like other linear approximations, the result is continuous and not differentiable. To address this, the `na.spline` function uses the cubic spline method to create a continuous and differentiable interpolation. The function `na.fill` sets the missing values to a single static value.

Other operations are also available. For instance, `na.omit` simply erases missing observations. The most powerful operation is `na.aggregate`, which allows the user to apply a function. For instance, setting the option FUN to `mean` fills all missing values with the arithmetic mean of the other values. This is interesting and appropriate in some cases, but time series often possess a trend that we might like to exploit. If we set the aggregator to a rolling mean, the missing value will lie on an appropriate trendline. The `zoo` package includes a function, `rollmean`, that provides this calculation. The results of some of these operations on the daily temperature data, along with the actual readings, are provided in Table 4.1.

Time series with missing data provide an excellent environment for testing one-dimensional interpolation systems. Time series occasionally are missing observations or we may wish to interpolate a value for an observation that might not normally be taken. Applying the interpolation in a meaningful way can be difficult. For instance, if a time series were rainfall, in centimeters per day, there is no reason to assume that just because it rained 2 centimeters one day and 4 centimeters two days later, that it rained 3 centimeters on the intervening day.

However, if it were a cumulative total, the interpolation may make sense. For instance, assume on the first day there were 2 centimeters of rain over the course of a month. Also assume that on the third, there were 13 centimeters total. It is reasonable to state that the value on the second day must be between 2 and 13, and this is even a requirement of the mean value theorem. Finding a result via interpolation is both possible and reasonable. Applying interpolation, though, creates new information. While that is clear in this example, it

	Source	LOCF	Linear	Spline	Mean	True
2015-06-01	30.00	30.00	30.00	30.00	30.00	30.00
2015-06-02	19.40	19.40	19.40	19.40	19.40	19.40
2015-06-03	17.80	17.80	17.80	17.80	17.80	17.80
2015-06-04	NA	17.80	20.30	19.13	29.25	18.90
2015-06-05	22.80	22.80	22.80	22.80	22.80	22.80
2015-06-06	26.70	26.70	26.70	26.70	26.70	26.70
2015-06-07	25.00	25.00	25.00	25.00	25.00	25.00
2015-06-08	31.10	31.10	31.10	31.10	31.10	31.10
2015-06-09	30.00	30.00	30.00	30.00	30.00	30.00
2015-06-10	29.40	29.40	29.40	29.40	29.40	29.40
2015-06-11	35.00	35.00	35.00	35.00	35.00	35.00
2015-06-12	35.00	35.00	35.00	35.00	35.00	35.00
2015-06-13	NA	35.00	33.35	33.06	29.25	31.70
2015-06-14	31.70	31.70	31.70	31.70	31.70	31.70
2015-06-15	32.20	32.20	32.20	32.20	32.20	32.20
2015-06-16	33.90	33.90	33.90	33.90	33.90	33.90
2015-06-17	29.40	29.40	29.40	29.40	29.40	29.40
2015-06-18	30.00	30.00	30.00	30.00	30.00	30.00
2015-06-19	30.00	30.00	30.00	30.00	30.00	30.00
2015-06-20	30.60	30.60	30.60	30.60	30.60	30.60
2015-06-21	NA	30.60	32.43	34.20	29.25	33.30
2015-06-22	NA	30.60	34.27	37.56	29.25	32.80
2015-06-23	36.10	36.10	36.10	36.10	36.10	36.10
2015-06-24	29.40	29.40	29.40	29.40	29.40	29.40
2015-06-25	28.30	28.30	28.30	28.30	28.30	28.30
2015-06-26	NA	28.30	28.30	28.24	29.25	26.10
2015-06-27	NA	28.30	28.30	27.76	29.25	22.80
2015-06-28	NA	28.30	28.30	27.56	29.25	25.60
2015-06-29	28.30	28.30	28.30	28.30	28.30	28.30
2015-06-30	30.60	30.60	30.60	30.60	30.60	30.60

Table 4.1
Time series interpolation examples

is not always and when working with computer graphics, interpolation plays a critical role in displaying and modifying images. This process is grounded on creating new images from old.

4.4.2 Computer Graphics

Interpolation is a key component of computer graphics. Almost every method presented in this chapter, and many more, can be applied to graphics one way or another. For this application section, we will apply the methods to resizing images. We will use PNG files to demonstrate this.

PNG stands for "Portable Network Graphics" and is a relatively common

```
resizeImageNN <- function(imx, width, height) {
    imx.dim <- dim(imx)
    layers <- imx.dim[3]
    w.scale <- width / (imx.dim[1] - 1)
    h.scale <- height / (imx.dim[2] - 1)

    ## Create a new array to store the new image
    nn <- array(0, c(height, width, layers))

    for(l in 1:layers)
        for(h in 0:height) {
            y <- round(h / h.scale) + 1
            for(w in 0:width) {
                x <- round(w / w.scale) + 1
                nn[h, w, l] <- imx[y, x, l]
            }
        }

    return(nn)
}
```

R Function 4.9
Image resizing via nearest neighbor

standard among image formats. There is a package, called **png**, which supports reading and writing PNG files. Inside the **png** package is a function, **readPNG** that will open and read a PNG file. The function itself returns a multidimensional array, which is what we want to work with. The array returned is $h \times w \times l$, where h is the height of the image in pixels and w is the width. The third dimension, l, is more complicated. For a full color image, there will be either three or four layers. If there are three layers, each represents the intensity of a different color. The first layer represents red; the second represents green; the third represents blue. These colors make up an RGB color palette. However, some may have a fourth layer. This is the alpha channel, representing how transparent the pixel is, usually called opacity.

In this book, which is not printed in full color, we will be working with grayscale images. Grayscale images include one channel or two channels. The first channel is how bright, or white, the pixel is, versus black. The second channel, if it exists, is an alpha channel. The pictures we will be working with have two channels.

In the first case, we will see how to apply the nearest neighbor algorithm to image interpolation. Given an image of some size, we may want to scale the image up or down to fit on a screen, or on this page, or for some other reason. Our first example is a 400×400 pixel image known as CKBH, shown in Figure

Figure 4.13
CKBH image at 400 pixels and 200 pixels, scaled using nearest neighbor

4.13.[1] Using the `resizeImageNN` function, shown in Function 4.9, we use the nearest neighbor function to resize the image to 200 × 200.

In grayscale, this is just the intensity channel and the alpha channel. If the image were in color, we would do this once for each channel. The `resizeImageNN` function itself returns an array, so we use `writePNG` to resave that file as a PNG. The reduced image is shown in Figure 4.13.

```
> library(png)
> ckbh <- readPNG("ckbh.png")
> ckbh200 <- resizeImageNN(ckbh, 100, 100)
> writePNG(ckbh200, "ckbh-200.png")
```

In this case, we have a clean reduction. We are eliminating three pixels for every one kept. A less generalized implementation could have cut the image in half in each dimension by eliminating every even row and every even column from the array. The results would be identical. The function `resizeImageNN` provides a simple generalized approach to image resizing using nearest neighbor.

When using nearest neighbor on reduced image dimensions, detail is lost. The data of the missing pixels is simply dropped, never to be seen again. In the example of CKBH, that is not a big concern. There is sufficient detail with one-quarter of the pixels to still understand the image and see a boy eating bread. If we wished to increase the image dimensions, we can, but we have no method of replacing the detail that is nonexistent in the data. Nearest neighbor will expand pixels, as necessary, to reach the desired image size. As

---

[1] CKBH are the initials of the author's son, and in this picture, he is eating a piece of bread in a park.

Figure 4.14
BABH image at 400 pixels and 200 pixels, scaled using bilinear interpolation

images grow progressively larger, the image may appear blocky or jagged, sometimes familiar to individuals printing images from the Internet.

We can use other methods to interpolate for better results. One other method is bilinear interpolation over the image. This works because an image is a collection observations, that is, pixel values, across a grid of observation points, the x and y coordinates. Since an image file is a rectilinear grid of data, bilinear interpolation is a suitable solution to resizing the image.

For this example, we will use a 400×400 pixel image known as BABH.[2] Using the `resizeImageBL`, shown in Function 4.10, we use bilinear interpolation to resize the BABH image to 200×200. Like `resizeImageNN`, `resizeImageBL` returns an array that we can resave as a new PNG. Unlike `resizeImageNN`, the function internally constructs an array that is passed to the `bilinear` function, rather than reimplementing the core method. The result is shown in Figure 4.14.

```
> babh <- readPNG("babh.png")
> babh200 <- resizeImageNN(babh, 100, 100)
> writePNG(babh200, "babh-200.png")
```

This reduction is very different from the nearest neighbor interpolation. Instead of just grabbing the nearest pixel, we create "new" information, of sorts. If we are reducing the size of the image, we create averages of the closest pixel values, weighted by the distance. The original information is lost. If we are scaling the image up, we create new pixels, where appropriate, based on the weighted average of the closest pixel values. Again, we create new information, and previous information may be lost.

---

[2]BABH are the initials of the author's daughter, and in this picture, she is standing on a park bench.

```
resizeImageBL <- function(imx, width, height) {
    imx.dim <- dim(imx)
    layers <- imx.dim[3]
    w.orig <- imx.dim[1]
    h.orig <- imx.dim[2]

    w.scale <- (w.orig - 1) / width
    h.scale <- (h.orig - 1) / height

    ## Create a new array to store the new image
    bl <- array(0, c(height, width, layers))

    zb <- matrix(0, 2, 2)
    for(l in 1:layers)
        for(h in 1:height) {
            y <- floor(h.scale * h)
            if(y == 0) y <- 1
            if(y == h.orig) y <- h.orig - 1

            yb <- c(y, y + 1)
            yl <- h.scale * h

            for(w in 1:width) {
                x <- floor(w.scale * w)
                if(x == 0) x <- 1
                if(x == w.orig) x <- w.orig - 1

                xb <- c(x, x + 1)
                xl <- w.scale * w

                zb[1, 1] <- imx[y, x, l]
                zb[1, 2] <- imx[y, x + 1, l]
                zb[2, 1] <- imx[y + 1, x, l]
                zb[2, 2] <- imx[y + 1, x + 1, l]

                bl[h, w, l] <- bilinear(xb, yb, zb, xl, yl)
            }
        }

    return(bl)
}
```

R Function 4.10
Image resizing via bilinear interpolation

There is a variety of options for interpolating images when rescaling. In addition to these methods, an approach known as bicubic interpolation creates a two-dimensional analog to the cubic spline function. Like the one-dimension version, rough edges and two-dimensional corners are smoothed resulting in cleaner images and fewer jagged edges. Another method, known as Lanczos resampling, weights neighboring points based on the sinc function. These methods are available in common image manipulations programs, including the freely available GIMP graphics package.

Comments

Interpolation methods provide a rich and diverse set of tools for solving missing data problems, but the missing data problem can take on many different forms. In some cases, we are interested in estimating a value where we know two endpoints. The intermediate value theorem, a staple of calculus, tells us that all values between the endpoints are taken on, though it does not limit the values taken on. If we make the conservative assumption that things proceed simply, then lower-degreed polynomial interpolation provides an excellent guide to estimated function values. Polynomial interpolation also bridges the last and next chapters in this book, as polynomial interpolation is implemented using linear algebra and an embedded interpolation provides the basis for numerical integration in the next chapter.

However, the restriction on lower-degreed polynomials may not suit the data given. For pure polynomial interpolation, the resultant function will have a degree of $n - 1$ where n is the number of points. For more points requiring interpolation, we need a higher-order polynomial, which we have already acknowledged is undesirable. So we have also walked through piecewise options, which increase the complexity of the process, while reducing the order of the solution.

For some problems, we are interested in analyzing the properties of the curve that generated the data we have. For these sorts of problems, polynomial, linear, and cubic spline interpolators are all suitable to one degree or another. Each results in a description of a generator that created our data, though the generator may not be very accurate. Depending on the type, we may be able to generate new data or we can analyze the generator as though it were the source function.

In addition to polynomials interpolations, including the Bézier curve, we are sometimes less interested in the structure of the underlying data than we are in filling in missing values. This helps when we want to analyze the underlying data, rather than its generator function. This is important when our goal is to use the completed dataset as part of a larger analysis, such as a linear model or time series analysis. As a result, many missing data methods have become staples of econometrics, statistics, and data science. Finally, we saw how we can use interpolation to resize images. Similar methods can be used to play compressed sound files smoothly or manage video streams.

Some, like nearest neighbor, are not particularly numerical in theme and implementation. Others, like spline interpolation, are key numerical methods. In this way, numerical analysis and data science share a focus on understanding the problem, its meaning, and its application. Together, these understandings lead us to the best solution for a problem. The wealth of interpolation methods make selection critically dependent on the application. Knowing how the results will be used can simplify selection and knowing the constraints of run time can support making better interpolation algorithm decisions.

Exercises

1. Given points $(3, 5)$, $(0, -2)$, and $(4, 1)$, find a quadratic function that interpolates the points using the Vandermonde matrix.

2. The Vandermonde matrix for finding polynomial interpolations solves the matrix to find the coefficients. Find the number of row operations necessary to interpolate a polynomial of degree n.

3. Create a function to evaluate a piecewise linear function, resulting from `pwiselinterp` for a given value of x.

4. Given $f(x) = x(x - 1)(x - 2)(x - 3)$, find the ordered pairs at the integers $x = 0, 1, 2, 3$. Find an interpolating polynomial. How does this compare to the graph of the actual equation?

5. The result of `pwiselinterp` is not well-suited for use in a real application. Rewrite the function to return a function that can calculate the results of the piecewise function for a given value of x.

6. Given 20 data points, in two dimensions, would we prefer to use a polynomial interpolation, or a spline? Why?

7. Given points $(3, 5)$, $(0, -2)$, and $(4, 1)$, find a Bézier curve that interpolates the points. Is this curve different from the Vandermonde polynomial interpolation of these points?

8. Use the recursive formula of the Bézier curve in equation 4.27 to create an R implementation of a generalized Bézier curve.

9. Given points, in three dimensions, $(1, 2, 1)$, $(4, 1, 3)$, and $(3, 1, 1)$, create a three-dimensional Bézier curve that interpolates these points.

10. Rewrite the `func:qbezier` function to operate on points in three dimensions.

11. Table 4.1 shows several approaches for filling in missing time series.

One, the average, uses the series mean to fill all missing values. Some applications may benefit from a moving average, over several nearby values. Write a function to implement a moving average interpolation for time series.

12. The nearest neighbor and bilinear applications to PNG files assumes the graphic is grayscale with one channel for intensity (whiteness) and one channel for alpha. Color graphics for a screen include three color channels with separate intensities for red, green, and blue, along with an alpha channel. Reimplement the nearest neighbor and bilinear applications for color PNG files. What is similar and what is different versus grayscale?

5

Differentiation and Integration

5.1 Numerical Differentiation

In analytical calculus, differentiation and integration are intrinsically linked by each being the inverse process of the other. So we might expect to see the strength of connection between their numerical counterparts. However, while numerical integration is a fully developed subfield of numerical analysis, with a wealth and range of methods available, numerical differentiation only offers a handful of closely related methods. Worse, those methods leave much to be desired.

In this section, we will first review the various finite difference methods. These methods use the definition of the derivative to estimate the derivative value. Finally, we will extend these methods to look at the second derivative of a function.

5.1.1 Finite Differences

Numerical differentiation, like other problems in numerical analysis, begins with the analytical definition of the problem. Differentiation is finding the slope of a line at a given point, and that discussion often starts with analysis that looks like,

$$f'(x) \approx \frac{f(x+h) - f(x)}{h}, \tag{5.1}$$

for some arbitrarily small value of h, the step size.

We probably recall the formal definition of the derivative is the limit of equation 5.1 as h goes to 0. So we want to make h as small as possible to get the best estimate of the derivative. We are limited to reasonable values, due to concerns about dividing by unusually small numbers. We must be additionally concerned about ensuring that $f(x)$ and $f(x+h)$ are sufficiently far apart, so there is no floating point round-off error affecting the intermediate subtraction.

Function 5.1, the `findiff` function, implements this finite differences method to differentiate a given function. The implementation is a single line and may be the simplest algorithm included in this book.

Applying this function to a linear equation should yield the coefficient of the x term, regardless of the step size. For the function $f(x) = 3x - 1$, this expectation holds at both small and large step sizes:

```
findiff <- function(f, x,
                h = x * sqrt(.Machine$double.eps)) {
    return((f(x + h) - f(x)) / h)
}
```

R Function 5.1
Finite differences formula

```
> f <- function(x) { 3 * x - 1 }
> findiff(f, 4, h = 1)
[1] 3
> findiff(f, 4, h = 1e-6)
[1] 3
```

which returns reassuring results of 3 for both step sizes. For a nonlinear function, the returns become estimates with a quality dependent upon the step size. For example, consider the Function $f(x) = \sin x$. From calculus, we know that the derivative of sin is cos and $f'(x) = \cos x$.

As shown in Figure 5.1, progressively smaller step sizes lead to estimates that more closely resemble the tangent line at $x = \pi/4$. We can see this in practice with the function findiff, where the derivative of $\sin x$ at $x = \pi/4$ is $\cos \pi/4 = \sqrt{2}/2 \approx 0.7071068$.

```
> findiff(sin, pi/4, h = 1)
[1] 0.2699545
> findiff(sin, pi/4, h = 0.5)
[1] 0.5048857
> findiff(sin, pi/4, h = 0.01)
[1] 0.7035595
```

In the findiff function, our algorithm finds the slope between x and $x+h$, called the forward difference. This is no better nor no worse than using the difference between x and $x - h$, commonly called the backwards difference. A negative step size, if passed to findiff, will be used. Intuitively, we know that this approximation improves as h approaches zero because of its relationship to the definition of the derivative. But we can move a step closer and also estimate the error of our derivative estimate. Taylor's theorem, which can approximate an nth differentiation of a function by an nth-order polynomial, states that the error term is,

$$f'(x) = \frac{f(x + h) - f(x)}{h} - \frac{h}{2}f''(c),\qquad(5.2)$$

where c is a value on the domain $[x, x + h]$. But the important part to understand is that the error term is governed by the size of h and the second derivative of f. As either decreases, so does the error. As the function is usually fixed by the application requirements, it is beyond our control. However,

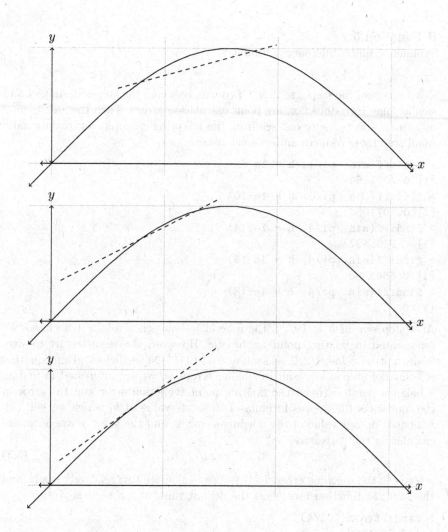

Figure 5.1
Derivative of $\sin x$ at $x = \pi/4$ for step sizes $h = \{1, 0.5, 0.01\}$

```
symdiff <- function(f, x,
              h = x * .Machine$double.eps^(1/3)) {
    return((f(x + h) - f(x - h)) / (2 * h))
}
```

R Function 5.2
Symmetric finite differences formula

we can control the step-size in h. Of course, reducing the step-size smaller and smaller may introduce floating point calculation errors. Then the results may be more suspect. As we can see from the following example, an exceptionally small step size produces an incorrect result:

```
> findiff(sin, pi/4, h = 1e-6)
[1] 0.7071064
> findiff(sin, pi/4, h = 1e-10)
[1] 0.7071066
> findiff(sin, pi/4, h = 1e-14)
[1] 0.7105427
> findiff(sin, pi/4, h = 1e-15)
[1] 0.6661338
> findiff(sin, pi/4, h = 1e-18)
[1] 0
```

At a step size of $h = 10^{-18}$, the difference between x and $x + h$ cannot be represented in floating point arithmetic. However, the results start to vary from the true value, $\sqrt{2}/2$, as soon as $h < 10^{-14}$. Nevertheless, given equation 5.2, smaller step sizes result in smaller errors, so we are interested in finding a balance point between the floating point trunction error and the error in the numerical differences formula. The "best" value of h, which we will call $h^\star$, based on our value of the machine error, ϵ, and the value x where we are calculating the derivative,

$$h^\star = x\sqrt{\epsilon}, \tag{5.3}$$

where ϵ is the machine error. In R, we can calculate this relatively easily, and the findiff function uses $h^\star$ as the default value of h if none is given.

```
> findiff(sin, pi/4)
[1] 0.7071068
> f <- function(x) { x^2 + 3 * x - 4 }
> findiff(f, 2)
[1] 7
```

Unfortunately, this leads to a value of $h^\star = 0$ at $x = 0$, and there is no reasonable way to estimate the best step size at $x = 0$.

From here, we can observe that another approach to finite differences is to use a version of the finite differences formula symmetric around x. This formula,

$$f'(x) = \frac{f(x+h) - f(x-h)}{2h} - \frac{h^2}{6}f'''(c), \qquad (5.4)$$

works in the same way as 5.1, but uses a step size twice as large, for a given h, and the difference domain is centered on x. The result is a single formula that takes the mean of the forward and backward difference methods, thereby reaching a more accurate result. This is called the central differences formula. As h can be arbitrarily small, subject to machine limits, multiplying h by 2 does not change anything. And like the finite differences formula, the error term is bounded, but is now governed by the size of h^2 and the third derivative of $f(x)$.

The function `symdiff`, given in Function 5.2 implements this algorithm. Using $f(x) = \sin x$ as the function again, we can see the results of the centered finite differences method:

```
> symdiff(sin, pi/4, h = 0.01)
[1] 0.707095
> symdiff(sin, pi/4, h = 0.001)
[1] 0.7071067
> symdiff(sin, pi/4, h = 0.0001)
[1] 0.7071068
> symdiff(sin, pi/4)
[1] 0.7071068
```

These results are quite a bit better. If the symmetric difference method were implemented in the same manner as the initial finite difference method, comparability would hold. In general, however, the symmetric finite differences formula will return more accurate results because it takes into account the function's behavior on either side of the examined x value.

Like the forward and backward difference formulae, the centered difference formula leads to a specific estimate of the best h value,

$$h^\star = x\sqrt[3]{\epsilon}, \qquad (5.5)$$

where ϵ is the machine error. In R, we can calculate this relatively easily, and the `symdiff` function uses $h^\star$ as the default value of h if none is given.

5.1.2 The Second Derivative

While it is useful to find the first derivative of a function, more derivatives are also possible using this method. The formula for the second derivative of $f(x)$ is,

$$f''(x) = \frac{f(x+h) - 2f(x) + f(x-h)}{h^2} - \frac{h^2}{12}f^{(4)}(c), \qquad (5.6)$$

where x, h, and c have the same meanings as before. This method is provided in Function 5.3, the `findiff2` function.

```
findiff2 <- function(f, x, h) {
    return((f(x + h) - 2 * f(x) + f(x - h)) / h^2)
}
```

R Function 5.3
Second derivative via finite differences

Using finite differences to estimate the second derivative of a function is, obviously, less accurate than finding the first derivative estimate. As we know, the second derivative of $\sin x$ is $-\sin x$, so the results of each of these should be similar.

```
> findiff2(sin, pi/4, h = 1e-4); -sin(pi/4)
[1] -0.7071068
[1] -0.7071068
> findiff2(sin, 3, h = 1e-4); -sin(3)
[1] -0.14112
[1] -0.14112
```

Indeed, these results are similar and are accurate to several decimal places. We can use this basic method of finite differences to find the third derivative, fourth derivative, and so on. The formula becomes more complicated at greater derivative levels and the error gets larger and larger, relative to the measurement itself.

5.2   Newton–Cotes Integration

Newton–Cotes integration forms rely upon the specification of analytical integration to be effective. Analytical integration is premised on the idea that the area under the curve can be partitioned into a number of subintervals which are regular and with well-understood area functions. Typically, this means we would divide the area under the curve into successively smaller rectangles of known width and height. The integral is defined as the limit of the sum of those spaces' areas as n approaches infinity and the width becomes infinitesimally small. This is the standard limit definition of the integral.

Newton–Cotes integration uses this definition and calculates the area under the curve by dividing it into rectangles, mirroring the definition of the integral. Though the definition of the integral implies an infinite number of infinitesimally small rectangles, R can provide a reasonable estimate of an integral using a limited number of rectangles. Subintervals, for the purposes of Newton–Cotes integration, are equally spaced and the Newton–Cotes process calculates an interpolated polynomial function representing the curve over the subinterval's width. Newton–Cotes integration increases in accuracy with both the number of subintervals and the degree of the interpolating polynomial.

Newton–Cotes integration forms work well with continuous integrands over a finite area. As the limits of integration define the number of subintervals of the area, an infinite bound is not computationally reasonable. If the integrand is discontinuous, Newton–Cotes integration may successfully evaluate, subject to the restriction that our results may not be usable. A jump discontinuity would evaluate successfully while an infinite discontinuity would evaluate provided no discontinuity lay on a subinterval boundary, including the limits of integration. A removable discontinuity may or may not successfully evaluate depending on the circumstances of the integrand. Like many other problems, graphing or performing some initial analysis will help us predict problems before we encounter them.

5.2.1 Multipanel Interpolation Rules

The midpoint rule, sometimes called the rectangle rule, and the trapezoidal rule are the two most basic examples of Newton–Cotes forms available, representing a one-point interpolation of the curve and a two-point interpolation of the curve, respectively. The midpoint rule provides the most elementary form, by dividing the integral range into m equally spaced subintervals. We treat each subinterval as a rectangle that is as wide as the subinterval, w, with a height equal to the integrated function at the midpoint of the interval, h. Of course, the area of each rectangle is $h \times w$. The sum of these rectangle areas is the approximated integral. Mathematically, we can state this as the Riemann sum,

$$\int_a^b f(x)\,\mathrm{d}x \approx \sum_{i=1}^m f\left(i\frac{|b-a|}{m} - \frac{|b-a|}{2m}\right)\frac{|b-a|}{m}. \tag{5.7}$$

The limit of the Riemann sum, as m approaches infinity, is the definition of the integral.

Equation 5.7 includes addition complexity in order to place the evaluation point at the midpoint of each subinterval. Simplified versions which evaluate at either the left or right endpoint of the subinterval are also possible. In practice, an algorithm would iterate over the subintervals, calculating the value of the function at that point representing the height of the rectangle, and summing the results before multiplying by each rectangle's width.

Figure 5.2 provides a graphical example of this approximation for equation 5.8 for $m = 2$, on the left In this example, the dashed rectangles represent the area that the midpoint method will calculate the area of. The solid line is the true integral. The rectangles include space that is not a part of the integral and exclude space that should be. Those spaces collectively are the error. On the right is the same example with $m = 4$. There are areas that are included and others that are excluded, but overall, the error portions are smaller. So is the total error.

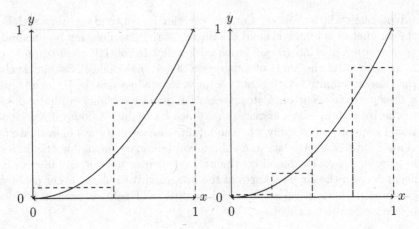

Figure 5.2
The midpoint rule for $f(x) = x^2$ with $m = 2$ and $m = 4$

For instance, consider the example of

$$\int_0^1 x^2 \, dx. \tag{5.8}$$

Solving this problem analytically is straightforward:

$$\int_0^1 x^2 \, dx = \left.\frac{x^3}{3}\right|_0^1 = \frac{1^3}{3} - \frac{0^3}{3} = \frac{1}{3}. \tag{5.9}$$

Since approximating the value of an integral requires n evaluations of the integrand using the midpoint method, estimating the value of equation 5.8 with $m = 2$ requires two evaluations at $x = \{0.25, 0.75\}$. Therefore,

$$\int_0^1 x^2 \, dx \approx (f(0.25) + f(0.75)) \frac{1 - 0}{2} = \frac{0.25^2 + 0.75^2}{2} = 0.3125. \tag{5.10}$$

Given the exact value of $1/3$ found in equation 5.9, this is a reasonable first approximation of the integral.

The R implementation of the midpoint rule is presented in Function 5.4, the `midpt` function. This function expects four arguments, including the function to integrate, the lower and upper bounds of integration, and the number of subintervals to use, defaulting to 100. Additional arguments are passed to the integrated function. The integrand shall accept vector arguments and shall return a vector of y values.

To use this function, we must define a function `f(x)` to represent the integrand and we can test the midpoint rule's results with increasing values of m:

```
midpt <- function(f, a, b, m = 100) {
    nwidth = (b - a) / m
    x = seq(a, b - nwidth, length.out = m) + nwidth / 2
    y = f(x)

    area = sum(y) * abs(b - a) / m
    return(area)
}
```

R Function 5.4
The midpoint rule

```
> f <- function(x) { x^2 }
> midpt(f, 0, 1, m = 2)
[1] 0.3125
> midpt(f, 0, 1, m = 10)
[1] 0.3325
> midpt(f, 0, 1, m = 100)
[1] 0.333325
> midpt(f, 0, 1, m = 1000)
[1] 0.3333332
```

As n increases, the approximation of the integral via the midpoint rule grows both more precise and more accurate. However, this increasing precision and accuracy comes at the cost of addition computing power necessary for the calculation. For a trivial integrand, like equation 5.8, this is not significant. More complex integrals will require more computing power to achieve less than desirable results.

This leads to the trapezoid rule, which uses trapezoids over the interval instead of rectangles. Consider the limit definition of the integral:

$$\int_a^b f(x)\,\mathrm{d}x = \lim_{m \to \infty} \sum_{i=1}^m \frac{(c_{i+1} - c_i) * (f(c_{i+1}) + f(c_i))}{m}, \tag{5.11}$$

where

$$c_i = a + \left(\frac{b-a}{n}\right) i, \tag{5.12}$$

and n is an arbitrary number of subintervals of the interval to be integrated over. This definition uses the trapezoid rule, implicitly, over increasingly smaller subintervals. Figure 5.3 shows the trapezoid rule for equation 5.8 for $m = 2$ on the left. Here, the dashed lines still show an overestimate of the integral value. Increasing the number of subintervals to 4 shows a dotted line nearly overprinting the true function values, clearly the closest estimate we have seen. Most integrals will not behave so well.

The R implementation of the trapezoid rule is presented in Function 5.5,

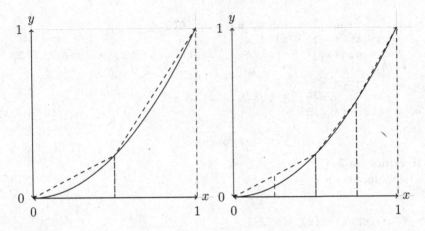

Figure 5.3
The trapezoid rule for $f(x) = x^2$ with $m = 2$ and $m = 4$

```
trap <- function(f, a, b, m = 100) {
    x = seq(a, b, length.out = m + 1)
    y = f(x)

    p.area = sum((y[2:(m+1)] + y[1:m]))
    p.area = p.area * abs(b - a) / (2 * m)
    return(p.area)
}
```

R Function 5.5
The trapezoid rule

the `trap` function. Like the midpoint rule implementation, this function expects four arguments, and otherwise behaves identically to the midpoint rule.

Each panel requires two evaluations of the integrand, but for the internal panels, the right endpoint of a panel is the left endpoint of the next. We need to evaluate each point only once, as shown in Function 5.5. Therefore, the trapezoid rule, for n panels, requires $n + 1$ evaluations of the integrand. When only two panels are used, the function needs to be evaluated at three points, a, $(b - a)/2$, and b. For Figure 5.3, $a = 1$ and $b = 0$:

$$\int_0^1 x^2 \, dx \approx \frac{(.5)(.25 + 1.25)}{2} = .375. \qquad (5.13)$$

That estimate is farther from the true value of $1/3$ than the midpoint estimate for $m = 2$. However, like the midpoint rule, the trapezoid rule approaches the correct result as m increases:

```
simp <- function(f, a, b, m = 100) {
    x.ends = seq(a, b, length.out = m + 1)
    y.ends = f(x.ends)
    x.mids = (x.ends[2:(m + 1)] - x.ends[1:m]) / 2 +
        x.ends[1:m]
    y.mids = f(x.mids)

    p.area = sum(y.ends[2:(m+1)] + 4 * y.mids[1:m] +
                    y.ends[1:m])
    p.area = p.area * abs(b - a) / (6 * m)
    return(p.area)
}
```

R Function 5.6
Simpson's rule

```
> f <- function(x) { x^2 }
> trap(f, 0, 1, m = 2)
[1] 0.375
> trap(f, 0, 1, m = 10)
[1] 0.335
> trap(f, 0, 1, m = 100)
[1] 0.33335
> trap(f, 0, 1, m = 1000)
[1] 0.3333335
```

The midpoint and trapezoid rules provide two basic methods for estimating the area under the curve, but both rely on straight line approximations for estimating the curve. For many purposes, this is sufficient and provides a reasonable approximation of the integral. However, higher-degreed interpolations of the curve yield more accurate approximations of the area under the curve, for the same number of panel evaluations.

For instance, interpolating a curve with a quadratic equation would provide a polynomial with a well-defined integral and only one additional function evaluation per panel, over the trapezoid rule. This integration method is known as Simpson's rule and an implementation is provided in Function 5.6, the simp function.

Further, a cubic interpolation of the integrand provides a polynomial of degree three requiring four evaluation points per panel. While this is a small number, it is still twice as many as the trapezoid rule. This method, called Simpson's 3/8, is implemented as simp38 in Function 5.7.

For a quadratic function, Simpson's rule will exactly interpolate the underlying function with only one panel, leading to an exact result with $m = 1$. Further, Simpson's 3/8 will not only exactly interpolate a quadratic function, it is more than necessary for such a function, though it is the minimally sufficient solution for a cubic function. Using the quadratic function, such as $f(x) = x^2$ used before, with only one panel over the domain, $[0, 1]$, we can see the results:

```
> midpt(f, 0, 1, m = 1)
[1] 0.25
> trap(f, 0, 1, m = 1)
[1] 0.5
> simp(f, 0, 1, m = 1)
[1] 0.3333333
> simp38(f, 0, 1, m = 1)
[1] 0.3333333
```

For integrating an example cubic function, $f(x) = x^3 - x^2 + 1$, with one panel over the domain $[0, 1]$, the true value is 2, shows growing convergence until the exact value is found by the function simp38:

```
> f <- function(x) { x^4 - x^2 + 1 }
> midpt(f, 0, 1, m = 1)
[1] 0.8125
> trap(f, 0, 1, m = 1)
[1] 1
> simp(f, 0, 1, m = 1)
[1] 0.875
> simp38(f, 0, 1, m = 1)
[1] 0.8703704
```

Of course, we are usually more interested in integrating functions that are not polynomials. For instance, trigonometric functions are not polynomial. Creating a reasonable approximation for them, that can be easily integrated, presents a better use case. Consider $f(x) = \sin x + \cos x$. Over the period from $[0, \pi]$, the integral of $f(x)$ is 2. Increasing the degree of the interpolating polynomial, shows we can move the approximation closer to the true value without getting there:

```
> f <- function(x) { sin(x) + cos(x) }
> midpt(f, 0, pi, m = 10)
[1] 2.008248
> trap(f, 0, pi, m = 10)
[1] 1.983524
> simp(f, 0, pi, m = 10)
[1] 2.000007
> simp38(f, 0, pi, m = 10)
[1] 2.000003
```

Like the midpoint and trapezoid rules, Simpson's rule and Simpson's 3/8 grow closer to the true value as the number of panels increases. Continuing with $f(x) = \sin x + \cos x$, using Simpson's rule:

```
simp38 <- function(f, a, b, m = 100) {
    x.ends = seq(a, b, length.out = m + 1)
    y.ends = f(x.ends)
    x.midh = (2 * x.ends[2:(m + 1)] + x.ends[1:m]) / 3
    x.midl = (x.ends[2:(m + 1)] + 2 * x.ends[1:m]) / 3
    y.midh = f(x.midh)
    y.midl = f(x.midl)

    p.area = sum(y.ends[2:(m+1)] + 3 * y.midh[1:m] + 3 *
                    y.midl[1:m] + y.ends[1:m])
    p.area = p.area * abs(b - a) / (8 * m)
    return(p.area)
}
```

R Function 5.7
Simpson's 3/8 rule

```
> simp(f, 0, pi, m = 2)
[1] 2.00456
> simp(f, 0, pi, m = 5)
[1] 2.00011
> simp(f, 0, pi, m = 10)
[1] 2.000007
> simp(f, 0, pi, m = 100)
[1] 2
```

And using Simpson's 3/8:

```
> simp38(f, 0, pi, m = 2)
[1] 2.00201
> simp38(f, 0, pi, m = 5)
[1] 2.000049
> simp38(f, 0, pi, m = 10)
[1] 2.000003
> simp38(f, 0, pi, m = 100)
[1] 2
```

By the time either method reaches 100 panels, the result is so close to 2, the result simply prints as 2. Analyzing this convergence process is the next step in understanding these functions and provides bounds on the errors of the estimates.

5.2.2 Newton–Cotes Errors

Newton–Cotes algorithms break the integrand into smaller and smaller subintervals, depending on the number of subdivisions specified. Each subinterval is fitted with an interpolating polynomial, which has a well-known and simply calculated integration function. The error in the integration is directly related to the error in the interpolating polynomial.

In the case of the trapezoid rule, the interpolating polynomial is a straight line between the two endpoints of the subinterval. For a linear equation, these straight lines would obviously match the linear equation. However, as shown in Figure 5.3, the interpolating polynomial can overestimate or underestimate the area of the function for even a simple quadratic equation. More complex functions provide more opportunity for the straight line of the trapezoid function to deviate from the true position of the analyzed function. The error of a Newton–Cotes integration is dependent upon the error of the interpolating polynomial in approximating the integrand.

Recall from Section 4.1.2, an interpolating polynomial will have a defined maximum error function,

$$f(x) = p(x) + h(x), \tag{5.14}$$

where $p(x)$ is the interpolating line has a maximum possible error of $h(x)$. From calculus, we remember

$$\int f(x)\,\mathrm{d}x = \int p(x) + h(x)\,\mathrm{d}x$$
$$= \int p(x)\,\mathrm{d}x + \int h(x)\,\mathrm{d}x, \tag{5.15}$$

and the upper limit of the error of the integration is $\int h(x)\,\mathrm{d}x$. The trapezoid rule finds the area under the curve of a one degree interpolating polynomial. Therefore, for a one panel integration,

$$\int_a^b h(x)\,\mathrm{d}x = -\frac{(b-a)^3}{12}f^{(2)}(\xi), \tag{5.16}$$

where ξ is some value between a and b, inclusive. For more panels, m, the error is found over each panel, and the aggregate error is the sum of the panel errors. For more than one panel, the error is,

$$\int_a^b h(x)\,\mathrm{d}x = -\frac{(b-a)^3}{12m^2}f^{(2)}(\xi), \tag{5.17}$$

where ξ again is a value between a and b.

We observe now that for any function with a second derivative of zero, the maximum error is also zero. That means for any straight line function, or lower degree polynomial, the trapezoid rule is precise, which makes sense. This logic extends to higher-order rules.

Increasing the number of interpolating points, and therefore the degree of the interpolating polynomial decreases the error. For Simpson's rule, which is a quadratic interpolation of three points, the error is,

$$\int_a^b h(x)\,\mathrm{d}x = -\frac{(b-a)^5}{180m^4}f^{(4)}(\xi), \tag{5.18}$$

and the error for Simpson's 3/8 is,

$$\int_a^b h(x)\,dx = -\frac{(b-a)^5}{80m^4}f^{(4)}(\xi). \tag{5.19}$$

Despite increasing the number of points, both Simpson's rule and Simpson's 3/8 lead to error terms including the fourth derivative of the function. For a third degree polynomial function, this is precise.

The midpoint rule, however, leads to a different conclusion. Despite only using one point, the midpoint formula is actually a quadratic interpolation, where the endpoints are not evaluated. Because of this, the error is actually smaller than for the trapezoid rule, where, for m panels, the error term is,

$$\int_a^b h(x)\,dx = -\frac{(b-a)^3}{24m^2}f^{(2)}(\xi), \tag{5.20}$$

which is half the size of the error for the trapezoid rule. We might find this surprising.

In the case of the rectangle method, we can evaluate the height of the rectangle at any point on the panel's interval. It is conceptually possible to evaluate at the left endpoint, the right endpoint, or anywhere in between, and implementations exist using all of them. Evaluating at the midpoint leads to a different type of Newton–Cotes algorithm, where the endpoints are not evaluated. These rules provide some advantages over others we have seen, so far.

5.2.3 Newton–Cotes Forms, Generally

Newton–Cotes forms share many common features by virtue of being based on interpolating polynomials. In their simplest forms, they use only one subdivision of the integration region. As Newton–Cotes forms are implemented in this section, this is as if $n = 1$, leading to only one panel. If a Newton–Cotes method uses more than one panel such that $n > 1$, these are called *composite* Newton–Cotes algorithms. As implemented in R, the composite form of the trapezoid method is not substantially more difficult than the equivalent noncomposite method, so separate implementations are not given.

It is also advantageous to increase the number of panels in a Newton–Cotes integration. Increasing the number of panels reduces the error bounds by reducing the potential error in the interpolation. This is shown in a number of examples in this section. But this advantage is limited. Increasing the number of panels increases the number of function evaluations necessary to calculate the integral. This increase is linear with respect to the number of panels selected.

Newton–Cotes forms also come in two broad classes. The first class is called *closed* Newton–Cotes forms. Closed Newton–Cotes forms are defined as those Newton–Cotes forms that evaluate the subject function, $f(x)$, at the endpoints of the integrand. The trapezoid and Simpson's rules are closed Newton–Cotes forms.

The second broad class is called *open* Newton–Cotes forms. Open Newton–Cotes forms do not evaluate the subject function at the endpoints of the integrand. The midpoint rule, which does not evaluate at the ends, meets this requirement and is an open Newton–Cotes method. This is advantageous if the function $f(x)$ does not have a value at the endpoint, a singularity. For instance, if $f(x) = \ln x$, then,

$$\int_0^1 \ln x \, dx = x \ln x - x. \tag{5.21}$$

Conveniently, over the region from 0 to 1, the value of the integral is -1. However, there is no value of $f(x)$ at $x = 0$. If integrated with the trapezoid rule or one of Simpson's rules, the integration will fail with a nonsensical value of negative infinity:

```
> trap(log, 0, 1, m = 10)
[1] -Inf
> simp(log, 0, 1, m = 10)
[1] -Inf
```

However, an open Newton–Cotes integration algorithm may estimate the value if the integrand is not evaluated at the singularity. The midpoint rule meets this requirement and is closer to the correct value as the number of subdivisions increases:

```
> midpt(f, 0, 1, m = 10)
[1] 1.301711
> midpt(f, 0, 1, m = 100)
[1] 1.301174
> midpt(f, 0, 1, m = 1000)
[1] 1.301169
```

But this can be misleading. For instance, let $f(x) = 1/x$. For this function, $\int f(x) \, dx = \ln x + c$ where c is the unknown constant. At $x = 0$, $\ln x$ is undefined. Accordingly, the result of $\int_0^1 f(x) \, dx$ is undefined. The trapezoid rule evaluates the integrand at the endpoints, and when evaluating at $x = 0$, fails:

```
> f <- function(x) { 1 / x }
> trap(f, 0, 1, m = 100)
[1] Inf
```

The value of "Inf" stems from the evaluation of $1/0$ in the `trap` function and clearly identifies a failure in the calculation. However, the midpoint rule will not evaluate the integrand at the endpoint and never fails. As n increases, the midpoint rule steadily progresses to infinity as the number of subdivisions used increases, without returning a hint of a problem:

```
> midpt(f, 0, 1, m = 10)
[1] 4.266511
> midpt(f, 0, 1, m = 100)
[1] 6.568684
> midpt(f, 0, 1, m = 1000)
[1] 8.871265
```

Selecting a Newton–Cotes form to use requires a basic understanding of the problem to solve. If the underlying integrand is undefined for any value of x, it is better to select a method that will avoid the discontinuity. It is also important to select the algorithm that will interpolate the integrand without unnecessary computational cost. The midpoint rule interpolates each panel using three points, but a rectangle rule using an endpoint uses one. The trapezoid rule uses two. Simpson's rule uses three and so on.

A Newton–Cotes integration algorithm can be developed using an arbitrarily large number of interpolating points per panel, n, and an arbitrarily large number of panels. The total number of function evaluations required is $m \times n$ and grows with each. A simpler algorithm, such as Simpson's rule, calculated a few thousand times can usually provide sufficiently accurate results for most purposes.

5.3   Gaussian Integration

Newton–Cotes methods are powerful, but they have two less than desirable features. One is that we must use $n + 1$ function evaluations for a result that is precise for a polynomial of degree n. This may seem like a good ratio, but in practice, that is $n + 1$ evaluations per panel, and we often dramatically increase the number of panels when evaluating integrals to increase precision. And increasing the number of panels will give us increased accuracy, but we usually will not gain precision.

The second drawback is less obvious. Newton–Cotes methods require the integrated function to be evaluated at equally spaced nodes. This is true regardless of function used. Each panel requires equally spaced nodes within them. This can be a problem with periodic functions, where periodic discontinuities can coincidentally land at evaluation points. If Newton–Cotes panels can cause problems, we may use Gaussian integration to solve an integral. Further, we gain increases in precision with Gaussian integration. In general, we reach a precision of a polynomial of degree $2n - 1$ for n function evaluations.

5.3.1   The Gaussian Method

Like Newton–Cotes integration, Gaussian integration is a weighted sum of function evaluations. While Newton–Cotes uses the results of a Vandermonde matrix, from Section 4.1.2, and its associated polynomial interpolation, we know we can find better approximations of a function. Gaussian integration relies on finding those better polynomial options. Where Newton–Cotes uses evenly spaced evaluation points with weights based on the polynomial interpolation of the points, Gaussian integration's weights and evaluation points are based on orthogonal polynomials.

Given $f(x)$, we know that for some function $w(x)$ producing weights, and $g(x)$ is a polynomial approximation to $f(x)$, then

$$\int_{-1}^{1} f(x)\, \mathrm{d}x = \int_{-1}^{1} w(x) g(x)\, \mathrm{d}x \approx \sum_{i=1}^{n} w_i f(x_i), \qquad (5.22)$$

where x_i is an evaluation point and w_i is the associated weight, and both are chosen to minimize error. Published weight and evaluation point lists are available in numerous books and online.[1]

In the elementary case, we start by setting $w(x) = 1$ and solving the equations so that,

$$w_1 x_1^i + \cdots + w_n x_n^i = \int_{-1}^{1} x^i\, \mathrm{d}x, \qquad (5.23)$$

for $i = 0, 1, \ldots, 2n - 1$, for the order of polynomial approximation desired. This value of $w(x)$ leads to the Legendre polynomials and we call this method Gauss–Legendre integration. We present the results for $n = 1, 2, 3, 4$ in Table 5.1. It is worth noting in the case where $n = 1$, the result is,

$$\int_{-1}^{1} f(x)\, \mathrm{d}x \approx 2f(0), \qquad (5.24)$$

which is equivalent to the midpoint rule introduced with the Newton–Cotes forms.

Other evaluation point and weight patterns can be developed using different values of $w(x)$. For instance, setting $w(x) = 1/\sqrt{1 - x^2}$ leads to the Chebyshev polynomials and Chebyshev–Gauss integration over $(-1, 1)$.

More interestingly, we can use different weight and evaluation point sets to evaluate integrals over infinite intervals. For instance, setting $w(x) = e^{-x}$ leads to the Gauss–Laguerre integration and it can be used to integrate over the interval, $[0, \infty)$. Gauss–Laguerre integration is appropriate for integration of the form,

$$\int_{0}^{\infty} w(x) f(x)\, \mathrm{d}x = \int_{0}^{\infty} e^{-x} f(x)\, \mathrm{d}x. \qquad (5.25)$$

We can also set $w(x) = e^{-x^2}$, leading to Gauss–Hermite integration over the

---

[1]For instance, Press et al. (2007) and Shampine, Allen, and Pruess (1997) provide such lists. Press et al. go further and provide code for generating weights and evaluation points in C++. In addition, lists of weights and points are available from Wikipedia. Standard reference values are also available from Abramowitz and Stegun (1964, ch. 25).

```
gaussint <- function(f, x, w) {
    y <- f(x)

    return(sum(y * w))
}
```

R Function 5.8
Gaussian integration method driver

interval, $(-\infty, \infty)$, leading to integration of the form,

$$\int_{-\infty}^{\infty} w(x)f(x)\,\mathrm{d}x = \int_{-\infty}^{\infty} e^{-x^2} f(x)\,\mathrm{d}x. \tag{5.26}$$

Integration points and weights are included in the cmna software package. Numerous other Gaussian integration evaluation point and weight sets are available for different integration intervals, beyond those provided here.

5.3.2 Implementation Details

The implementation of Gaussian integration is given in Function 5.8, the gaussint function. However, this implementation is little more than a driver for connecting integration points with weights. That is, the gaussint function requires as options both the evaluation points and their associated weights. Like the Newton–Cotes implementations, the integrand shall accept vector arguments and shall return a vector of y values.

Accordingly, for the simplest case of Gauss–Legendre integration, weights are given in Table 5.1 for up to $n = 5$. We can show the method in action for $n = 2$ and over the function, $f(x) = x^3 + x + 1$, along with a midpoint rule with only two function evaluations, for reference.

```
> w = c(1, 1)
> x = c(-1 / sqrt(3), 1 / sqrt(3))
> f <- function(x) { x^3 + x + 1 }
> gaussint(f, x, w)
[1] 2
> trap(f, -1, 1, m = 1)
[1] 2
```

For this cubic formula, both the Gaussian approach and the trapezoid method yield the same correct answer. Integrating the cosine function over the interval $[-1, 1]$ shows the differences better.

```
> gaussint(cos, x, w)
[1] 1.675824
> trap(cos, -1, 1, m = 1)
```

n	weights (w_i)	points (x_i)
1	2	0
2	1	$\pm\sqrt{1/3}$
3	8/9	0
	5/9	$\pm\sqrt{3/5}$
4	$\frac{18+\sqrt{30}}{36}$	$\pm\sqrt{\frac{3}{7}-\frac{2}{7}\sqrt{\frac{6}{5}}}$
	$\frac{18-\sqrt{30}}{36}$	$\pm\sqrt{\frac{3}{7}+\frac{2}{7}\sqrt{\frac{6}{5}}}$

Table 5.1
Gauss–Legendre points and weights for $n = 1, 2, 3, 4$ (Kythe and Schäferkotter 2005, 116)

```
[1] 1.080605
```

The true value is approximately 1.68. Gaussian integration and Newton–Cotes integration, using two evaluation points, yield very different answers, with Gaussian providing a more accurate result. However, in general, Gaussian integration is not necessarily better. The advantages are only possible when the integrand can be very closely approximated by a polynomial.

With the cmna package, there are gauss.legendre, gauss.laguerre, and gauss.hermite convenience functions. These each accept a parameter f that is the function to integrate and the parameter m which is the number of evaluation points. These convenience functions automatically load the correct evaluation point and weight set if available and call the gaussint driver function, returning the result. For instance,

```
> gauss.legendre(cos, 5)
[1] 1.682942
```

The function gauss.legendre is provided in Function 5.9, and the implementations of gauss.laguerre and gauss.hermite are essentially identical. The function will produce an error if the desired weight set is not available.

The gauss.legendre function only allows for integrating over the interval from $[-1, 1]$. It is possible to change the integration interval analytically, by observing that,

$$\int_a^b f(x)\,\mathrm{d}x = \frac{b-a}{2}\int_{-1}^1 f\left(\frac{a+b+u(b-a)}{2}\right)\,\mathrm{d}u. \qquad (5.27)$$

Using equation 5.27, an automated process can rescale the bounds of integration and the final result to deliver the result over an arbitrary interval from

```
gauss.legendre <- function(f, m = 5) {
    p <- paste("gauss.legendre.", m, sep = "")
    params <- eval(parse(text = p))

    return(gaussint(f, params$x, params$w))
}
```

R Function 5.9
Gauss–Legendre integration using precomputed weights

$[a, b]$. The gauss.laguerre function integrates over the region $[0, \infty)$ and the gauss.hermite function integrates over $(-\infty, \infty)$.

Because some Gaussian weight sets support integration over infinite intervals, the Gaussian family of methods is powerful enough to function on many integrands that Newton–Cotes methods cannot. However, the Gaussian family is more difficult to work with without precomputed weights and evaluation points. If precomputed points are available, then their use is straightforward. Their principal benefits are the lack of equally spaced evaluation points and the ability to select evaluation points that avoid singularities of the integrand. However, in the general case of a definite integral, we are likely better off working with an adaptive method that exploits the underling power of Newton–Cotes based techniques, as we will see in the next section.

5.4 More Techniques

For integration, we have seen two families of integration routines. In this section, we start looking at more complicated integrators. Initially, we are going to look at two kinds of extensions of Newton–Cotes algorithms. These expand Newton–Cotes plans and attempt to add some intelligence to their work. Adaptive routines exploit some fundamental assumptions about both the integral they are operating on and the algorithms for integration to add that intelligence. The Romberg integrator will attempt to use the known features of the error to find a better integral estimate from just a few iterations of the trapezoid rule. Finally, we will take a high-level view of Monte Carlo integration and applications.

5.4.1 Adaptive Integrators

Adaptive integration provides a different approach to estimating integral values. One of the central tenents of numerical analysis, and this book in general, is that we should commit to some sort of human analysis of a problem before attempting to solve it algorithmically. Numerical analysis methods are generally unable to solve all problems very well. So knowing something about the problem can allow us to better select numerical methods appropriate to the problem. For instance, in the context of numerical integration, a discontinuity at an endpoint would not be suitable for a closed Newton–Cotes solution.

Of course, it would be nicer if we could cajole the computer into learning something about the problem, instead of doing that work ourselves. Adaptive methods provide an approach to this. Adaptive integration methods examine the integral they are operating upon and change their own parameters to improve the quality of the integration. The most simple adaptive algorithms provide a brute force approach to quality improvement by examining the error of the integration. On the other hand, if we knew the error, we could theoretically just fix the estimate. That is where the error limits on the Newton–Cotes algorithms can help.

If we can find out something about the error, we can use that information to refine our estimate. Imagine we are integrating a function, $f(x)$ over $[a, b]$. If we used the midpoint rule, from Section 5.2.1, we know the maximum possible error is,

$$\int_a^b h(x)\,\mathrm{d}x = -\frac{(b-a)^3}{24m^2}f^{(2)}(\xi). \qquad (5.28)$$

Two observations immediately become apparent. Regardless of what the function $f(x)$ is, or its second derivative, two changes can be made to the integration to improve its quality. First, the error is proportion to the cube of the length of the integration domain. Reducing the length improves quality and cutting the length in half provides an eight-fold improvement in quality. Second, error is inversely proportional to the number of panels used, m. Increasing the panels reduces error and doubling the number of panels provides four-fold improvement in quality.

We can design an algorithm around these observations. First, we can estimate the value of the integral, Q_1, using a 1-point midpoint rule. Second, we can estimate it again using a 2-point midpoint rule, Q_2. Since we have doubled the number of points in the rule, we now know that the maximum difference between Q_1 and Q, the true value of the integral, is no more than 4 times further than the maximum difference between Q_2 and Q. And if $Q_2 - Q$ is less than some specified tolerance, then the difference between Q_1 and Q_2 must be less than three times the same tolerance.

We check the difference between the two estimates, and if the difference is greater than the tolerance, we may still be within the tolerance, but we are not certain yet, and probably are not. So the process divides the integration region in half, and applies the adaptive integrator to both halves, separately, summing the result.

The function `adaptint`, given in Function 5.10, implements this process, recursively. The implementation also includes a trap door, the parameter n that is decremented with each successive call to `adaptint`. This prevents the program from spinning out of control and never returning. If the maximum

```
adaptint <- function(f, a, b, n = 10, tol = 1e-6) {
    if(n == 1)
        area <- midpt(f, a, b, m = 2)
    else {
        q1 <- midpt(f, a, b, m = 1)
        q2 <- midpt(f, a, b, m = 2)
        if(abs(q1 - q2) > 3 * tol) {
            n = n - 1
            tol <- tol / 2
            c <- (a + b) / 2
            lt <- adaptint(f, a, c, n = n, tol = tol)
            rt <- adaptint(f, c, b, n = n, tol = tol)
            area <- lt + rt
        } else
            area <- q2
    }

    return(area)
}
```

R Function 5.10
Adaptive method with midpoint rule

recursive depth is reached, `adaptint` returns whatever value it can calculate without regard to its estimated error.

In addition, as the integration region is successively halved, so is the tolerance for the smaller regions. This ensures that the total possible error remains under the initially specified maximum tolerance. We can specify either the tolerance or maximum recursion depth, with reasonable defaults available for each.

The recursive depth here is only a maximum. If on a partial region, `adaptint` finds a region where the two-point midpoint rule yields a sufficiently accurate result, within the revised tolerance for that region, the function will go no further and return the two-point midpoint rule result for subsequent aggregation. This is the adaptive part in action.

In the worst-case scenario, the adaptive integrator will require a full reevaluation for every split. This leads to 2^n midpoint evaluations, which grows exponentially. In general, the adaptive integrator will find a sufficient result before reaching that point. However, each recursive call is an opportunity for the adaptive integrator to fail, split, and execute the recursive call again.

In this example, we integrate $f(x) = \sin^2 x + \ln x$ over the region from $[1, 10]$ using successively larger maximum recursion depth.

```
> f <- function(x) { sin(x)^2 + log(x) }
> adaptint(f, 1, 10, n = 4)
[1] 18.53653
```

```
> adaptint(f, 1, 10, n = 5)
[1] 18.52788
> adaptint(f, 1, 10, n = 6)
[1] 18.52568
```

And we can integrate the same function by focusing, instead, on the minimum error tolerance.

```
> f <- function(x) { sin(x)^2 + log(x) }
> adaptint(f, 1, 10, tol = 1e-3)
[1] 18.52491
> adaptint(f, 1, 10, tol = 1e-6)
[1] 18.52494
> adaptint(f, 1, 10, tol = 1e-9)
[1] 18.52494
```

The midpoint rule is the workhorse of the adaptive method here and could easily be replaced by a trapezoid rule, Simpson's rule, or one of the Gaussian rules, if appropriate. Some implementations use the trapezoid rule for the less accurate measurement and Simpson's rule for the more accurate measurement. The key to selecting measurements in an adaptive integrator is understanding the relative difference in errors and measuring the tolerance bounds, appropriately.

The adaptive integrator can break down in two important ways. In the first, we may see both the lower and higher tolerance estimates be close to each other, initially. That is, if the two estimates are initially within the tolerance, the integrator will happily return the value found, without regard to how close the estimate is to the actual value. Some implementors address this by refusing to return any result in the first n_* iterations, where n_* is some minimum number of iterations. The result helps force some measure of convergence.

In the opposite extreme, it is entirely possible Q_2 is within the tolerance of the true value, but $Q_2 - Q_1$ can be up to five times the tolerance apart. This will happen when both estimates are at opposite extremes of the error bounds. In this case, the integrator will continue executing to reduce the spread to within the specified level. In practice, this only yields a more accurate estimate, though the integrator will have no way of knowing this and cannot, therefore, inform us.

5.4.2   Romberg's Method

Like the adaptive integration algorithm, `adaptint`, Romberg integration is a relatively straightforward extension of the Newton–Cotes algorithm family. Both work by using refined iterations of some underlying Newton–Cotes method to provide a more accurate estimate of the integral's value. Unlike `adaptint`, Romberg integration is not an adaptive approach to integration. That is, it does not change its own behavior based on the behavior of the function to be integrated. Instead, we exploit the behavior of the trapezoid function at the limit to produce an integral estimate.

To understand Romberg integration, we must start with a recursive implementation of the trapezoid rule. If we start with a function, $\mathbf{T}(f, m)$ where T is the trapezoid function, f is a function to be integrated, and m is the number of panels to integrate over, then,

$$\mathbf{S}(f, m) = \frac{4\mathbf{T}(f, m) - \mathbf{T}(f, m/2)}{3}. \tag{5.29}$$

where $\mathbf{S}$ is Simpson's rule. Then, if we define $T(f, 0) = (b - a)(f(b) + f(a))/2$, then our recursive function is complete. Because of this relationship, the fraction given in equation 5.29 is also an approximation for the integral. Because this relationship is recursive, we can refine the approximation by including any depth of recursive iterations of the value.

In general,

$$\mathbf{I}_{j,k} = \frac{4^k \mathbf{I}_{j,k-1} - \mathbf{I}_{j-1,k-1}}{4^k - 1}, \tag{5.30}$$

where $\mathbf{I}_{0,0}$ is a one paneled trapezoid rule over the integral and $\mathbf{I}_{j,0}$ is the trapezoid rule with 2^j panels over the integral. Using these baseline functions, $\mathbf{I}_{j,k}$ can be found iteratively as a lower-triangular matrix where each not in the left-most column is a function of the value to the left and the entry above it.

This recursive definition emerges from Richardson extrapolation. When applied to the trapezoid algorithm, which converges to the true value of the integral as m, the number of panels, increases, the relationship in equation 5.30 emerges. It is important to understand that at the limit as k approaches infinity, the value of $\mathbf{I}_{j,k}$ is the true value of integral. For smaller values of k, the Romberg integral is still only an approximation, albeit a very good one.

The romberg function, given in Function 5.11, implements this algorithm. iteratively. The function accepts one value, m, in addition to the integral parameters that describes how many iterations to accomplish. After initially creating a matrix of size m, containing only the value NA, the lower-triangular matrix is developed from the baseline values if $\mathbf{I}_{j,0}$. Of course, the matrix starts with $\mathbf{I}_{1,1}$ because R matrices use 1 as the starting index. Importantly, this implementation is not given recursively, despite the core equation being defined recursively. This was done for programmatic flexibility. Other methods may choose to implement Romberg integration recursively.

Using the same sample function, $f(x) = \sin^2 x + \log x$ used to demonstrate adaptint, we can see the results of Romberg integration converge rapidly, with even a small number of iterations. Here, we can see the romberg function converge to a suitable precision after only 10 iterations:

```
> f <- function(x) { sin(x)^2 + log(x) }
> romberg(f, 1, 10, m = 3)
[1] 18.48497
```

```
romberg <- function(f, a, b, m, tab = FALSE) {
    R <- matrix(NA, nrow = m, ncol = m)

    R[1, 1] <- trap(f, a, b, m = 1)
    for(j in 2:m) {
        R[j, 1] <- trap(f, a, b, m = 2^(j - 1))
        for(k in 2:j) {
            k4 <- 4^(k - 1)
            R[j, k] <- k4 * R[j, k - 1] - R[j - 1, k - 1]
            R[j, k] <- R[j, k] / (k4 - 1)
        }
    }

    if(tab == TRUE)
        return(R)
    return(R[m, m])
}
```

R Function 5.11
Romberg's method for integration

```
> romberg(f, 1, 10, m = 5)
[1] 18.52473
> romberg(f, 1, 10, m = 10)
[1] 18.52494
```

This implementation of the Romberg integrator includes an additional option, *tab*, which defaults to FALSE. If set to TRUE, the function returns the entire table of integral estimates, rather than the best estimate found. Using the function $f(x) = \sin x$ as an example:

```
> romberg(sin, 0, pi, m = 5, tab = TRUE)
              [,1]      [,2]      [,3]      [,4]  [,5]
[1,] 1.923671e-16        NA        NA        NA    NA
[2,] 1.570796e+00  2.094395        NA        NA    NA
[3,] 1.896119e+00  2.004560  1.998571        NA    NA
[4,] 1.974232e+00  2.000269  1.999983  2.000006    NA
[5,] 1.993570e+00  2.000017  2.000000  2.000000     2
```

This tableau of results is remarkable. The initial estimate of the value of the integral $\sin x$ from 0 to π using a one-panel trapezoid method should be 0 and the upper left corner of the tableau is very close to 0. But with only 5 iterations of the algorithm, after roundoff error, the correct value of 2 has emerged. Further, the result was accurate to within six places after the fourth iteration. The speed of convergence for the Romberg integral makes a solid go-to method for integration when handling integrals over finite domains.

Using the trapezoid rule as the underlying method for Romberg integration can introduce the same problems as using the trapezoid method, in general. The most important of these is that the function must be evaluatable at a number of equally spaced points. Critically, this also includes the endpoints, which cannot be guaranteed for a number of interesting integrals. Concerns about the endpoints can be alleviated by using the midpoint rule as the underlying method. Some implementations of Romberg integration have been developed that are adaptive in response to discontinuities, or which pass off processing to a method better suited to the integral in question.

Finally, because each successive iteration improves the accuracy and precision of the estimate, and previous estimates are available in the tableau, this function could have a tolerance specification added to the parameters list. Such an implementation would successively check, like in the `adaptint` function, after each iteration the estimated error and stop processing if the given tolerance requirements were met.

5.4.3   Monte Carlo Methods

Monte Carlo methods have a long history in simulation and other areas of applied mathematics. Economists use Monte Carlo methods to model potential outcomes of the economy and physicists use Monte Carlo methods to determine the likely outcome of stochastic physical processes. Finance, engineering, and many other disciplines have adopted Monte Carlo methods to better understand random processes in the field.

The name Monte Carlo itself comes from the area in Monaco, famous for its casinos and gambling. Obviously, a good casino game is dependent upon randomness, as are Monte Carlo methods. The name illustrates the importance of randomness in the process as Monte Carlo algorithms use a random number generator to differentiate the input to a function.

The random number must come from the expected domain of the function. Further, the function itself is deterministic in that for given two inputs from the domain of the function, x_1 and x_2, if $x_1 = x_2$, then $f(x_1) = f(x_2)$. This establishes the repeatability of the function. The random number generator is used to generate a large number of inputs and the function is run over each input. Finally, the results are aggregated according to a logic model appropriate to the analysis being performed.

Monte Carlo methods can be used for numerical integration in any arbitrary number of dimensions. The fundamental approach is to place some number of points, m, randomly over the domain to integrate over. If the point lies "below" the function line, it is considered in the area of integration. If the point is "above" the function line, it is not. The area under the curve estimate is the percentage of points below the line.

Some of the earliest Monte Carlo algorithms were used to find the area under curves or to estimate the value of π, a favorite hobby of mathematicians. One approach creates a quarter circle, using the function $f(x) = \sqrt{1 - x^2}$.

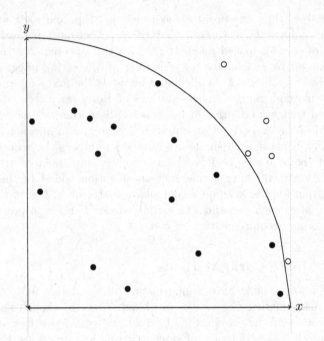

Figure 5.4
The Monte Carlo method for $f(x) = \sqrt{1 - x^2}$ with $m = 20$

Over the domain, $[0, 1)$, this is a function and is the result of solving the standard equation for the circle, $x^2 + y^2 = r$ for y where $r = 1$.

Figure 5.4 shows the plot of this function. In addition, a random 20 points are selected. If the point is below the curve, or inside the quarter circle, the point is solid and those points above the point are hollow. In this example, 15 of the 20 points are solid, leading to an area estimate of 0.75 over the domain. Because this represents a quarter of the circle, the estimate for π is 3, which is at least in the right neighborhood. Increasing the number of random points tests increases the precision of the estimate and the accuracy, on average.

Formally, this leads to a definition of Monte Carlo integration in one dimension. Here,

$$\int_a^b f(x)\,\mathrm{d}x \approx (b - a)\frac{1}{N}\sum_{i=1}^{N} f(x_i), \tag{5.31}$$

where N is the number of points the function is evaluating. The `mcint` function, Function 5.12, implements the Monte Carlo integration algorithm for integration in one dimension. We can see the increasing accuracy of the estimate for π as the number of evaluation points increases by powers of 10, from 100 to 1 million:

```
mcint <- function(f, a, b, m = 1000) {
  x <- runif(m, min = a, max = b)

  y.hat <- f(x)
  area <- (b - a) * sum(y.hat) / m
  return(area)
}
```

R Function 5.12
Univariate Monte Carlo integration

```
> f <- function(x) { sqrt(1 - x^2) }
> mcint(f, 0, 1, m = 1e3) * 4
[1] 3.185337
> mcint(f, 0, 1, m = 1e4) * 4
[1] 3.139844
> mcint(f, 0, 1, m = 1e5) * 4
[1] 3.14235
> mcint(f, 0, 1, m = 1e6) * 4
[1] 3.14187
```

However, even the estimate for 1 million points is only good for a few digits of π. And, for each point evaluated, the underlying function must also be evaluated. Compared to Newton–Cotes methods from Section 5.2, this is extremely expensive, for a comparatively low-quality result.

The advantage of using Monte Carlo methods stems from their ability to handle multiple integrals. Consider the example,

$$\int_0^1 \int_0^1 x^2 y \, \mathrm{d}x \, \mathrm{d}y. \tag{5.32}$$

The value of this integral is $1/6$. But Newton–Cotes methods we have seen will not get us there. Monte Carlo can and an example implementation of Monte Carlo implementation, over two variables, is provided in Function 5.13, the function `mcint2`. This is based on the two-dimensional extension of equation 5.31:

$$\int \int f(x,y) \, \mathrm{d}x \approx V \frac{1}{N} \sum_{i=1}^{N} f(x_i, y_i), \tag{5.33}$$

where V is the area of the x-y cross-section that $f(x,y)$ is integrated over.

This function works analogously to `mcint`, except the domain for x and the domain for y are set via arrays. Otherwise, the function is a straightforward bivariate extension of `mcint`. We can see the results on the integral in equation 5.32:

```
> f <- function(x, y) { x^2 * y }
> mcint2(f, c(0, 1), c(0,1), m = 1e3)
```

```
mcint2 <- function(f, xdom, ydom, m = 1000) {
    xmin <- min(xdom)
    xmax <- max(xdom)
    ymin <- min(ydom)
    ymax <- max(ydom)

    x <- runif(m, min = xmin, max = xmax)
    y <- runif(m, min = ymin, max = ymax)

    z.hat <- f(x, y)
    V = (xmax - xmin) * (ymax - ymin)
    volume = V * sum(z.hat) / m
    return(volume)
}
```

R Function 5.13
Bivariate Monte Carlo integration

```
[1] 0.16242
> mcint2(f, c(0, 1), c(0,1), m = 1e4)
[1] 0.1646874
> mcint2(f, c(0, 1), c(0,1), m = 1e5)
[1] 0.1669154
> mcint2(f, c(0, 1), c(0,1), m = 1e6)
[1] 0.1668718
```

Like other examples, the integral gets closer to the true value as we increase the number of iterations used. However, the increase in dimensionality decreases the effectiveness of each increase of point estimates. However, it is not necessarily going to get closer to the true value, as the result is not deterministic.

```
> mcint2(f, c(0, 1), c(0,1), m = 1e4)
[1] 0.1622139
> mcint2(f, c(0, 1), c(0,1), m = 1e4)
[1] 0.1687029
> mcint2(f, c(0, 1), c(0,1), m = 1e4)
[1] 0.1652631
> mcint2(f, c(0, 1), c(0,1), m = 1e4)
[1] 0.1675628
```

Because it is nondeterministic, the error of Monte Carlo integration is not bounded in the sense we have seen so far. However, we can estimate the variance of the estimate produced, which decreases as the number of points increases:

$$\text{Var} \frac{1}{N} \sum f() = \frac{\sigma^2}{N}, \tag{5.34}$$

where $\sigma^2 = \text{Var } f()$. This holds for all dimensions. Additionally, in one and two dimensions, Monte Carlo is a relatively simple affair. Fortunately, the method can be trivially extended to higher dimensions.

The implementation of both `mcint` and `mcint2` are flawed in that they cannot integrate over regions where the value is negative. These implementations presume the lowest value of the dependent variables is 0, and only allow for setting a maximum value. This is not inherent in Monte Carlo approaches, but the streamlined implementation overestimates the integral's value by not subtracting negative regions. This is a correctable problem.

Monte Carlo integration methods provide another advantage in that they are trivially parallelizable, a topic beyond the scope of our discussion in this book. It is possible to have multiple computers execute Monte Carlo integration processes and aggregate the results, increasing the power, without substantial increase in wall clock time, the amount of time it takes in real-world measurements instead of computer time.

5.5 Applications

There are numerous applications of the integral throughout science, engineering, and mathematics. Integrals describe the work necessary to complete a task in physics, to find the cumulative distribution function in statistics, and to find the charge in electrical engineering. The number and diversity of these applications are a testament to the power of calculus. We will only look at two to show a pair of important methods.

The first focuses on the volume of solids created by revolving the plot of a function around one of the axes. While the implementations themselves are just a few lines, they illustrate the power of R with integrals to show how you can embed one R function within another. Many applications will rely on this method and understanding in a three-line example provides the outline for more complex applications.

The second is the Gini coefficient. The Gini coefficient is a measurement of income inequality for a population. Given the income data for quintiles of a population, we can use Newton–Cotes methods to estimate the Gini coefficient. With many populations, quintile data is available and our estimates are powerful enough for policymaking purposes.

5.5.1 Revolved Volumes

Many calculus textbooks provide a selection of applications of the integral after the basic integration methods have been introduced. The earliest are disconnected from the physical world but continue the theme of abstraction. Two of those abstract methods focus on the volume of solids of revolution.

And these provide an illustration of how we can implement complex functions in R.

Solids of revolution are the result of revolving a curve around an axis and tracing out the solid that is created. This is the abstract analogue of clay on a pottery wheel. As the wheel turns and the clay is shaped, a rotationally symmetric solid is spun. The solid of revolution mirrors this process.

In the first case, we choose to revolve the curve around the x-axis. Imagine a constant function were revolved around the x-axis. The result would be a solid defined by a straight line. This is a cylinder, as shown in Figure 5.5. The volume of a cylinder is $V = h\pi r^2$, where h is the height of the cylinder and r is the radius. This is the result of finding the area of a slice of the cylinder and multiplying by the height. The disc method extends this to any solid defined by any shaped curve. Accordingly, the volume of this revolved solid is,

$$V = \pi \int_a^b f(x)^2 \, dx. \tag{5.35}$$

The function `discmethod`, shown in Function 5.14, implements this formula.

This function works differently from a number of others we have created. Internally, a new function is created that returns the result of $\pi f(x)^2$. This new function, called `solid`, is the function of integration processed by the midpoint method, though any suitable integration method would suffice.

We can immediately test this by ensuring the results are correct for the integration of a constant function, so that $f(x) = 1$, which should return the volume of π, over the region of $[0, 1]$.

```
> ## f(x) is vector-safe with rep()
> f <- function(x) { return(rep(1, length(x))) }
> discmethod(f, 0, 1)
[1] 3.141593
```

Having reassured ourselves the `discmethod` function works as advertised, we can examine a more interesting solid. Figure 5.6 shows the results, in three dimensions, of revolving the function $f(x) = x^2$ over the region from $[1, 2]$. The resulting image resembles a cone, but the sides are curved, squared, in fact, and the narrow end has been cut off.

As we already know the function describing the curve, finding the volume with the disc method is:

```
> f <- function(x) { return(x^2) }
> discmethod(f, 1, 2)
[1] 19.47751
```

This sort of method has applications in 3D printing and other types of engineering. If a space can be described as a solid of revolution, we can calculate its volume using a three-line function combined with any integration algorithm.

We might also choose to revolve a function around the y-axis. In this case,

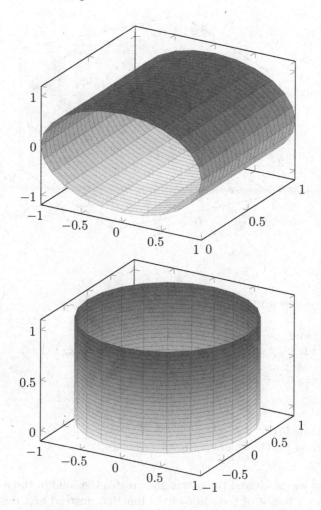

Figure 5.5
Cylinder as a rotated solid around the x-axis and y-axis

```
discmethod <- function(f, a, b) {
    solid <- function(x) { return(pi * (f(x))^2) }

    return(midpt(solid, a, b))
}
```

R Function 5.14
Disc method for the volume of revolved solids

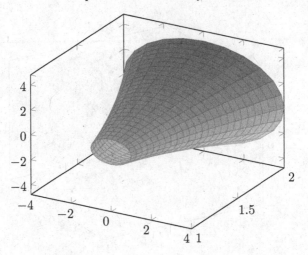

Figure 5.6
Rotated volume around the x-axis

```
shellmethod <- function(f, a, b) {
    solid <- function(x) { return(x * f(x)) }

    return(2 * pi * trap(solid, a, b))
}
```

R Function 5.15
Shell method for the volume of revolved solids

a function wraps around the y-axis and creates the solid in three-dimensional space with a height of the value of the function, instead of a radius with the value of the function. In this instance, the radius of the solid is described by the integration bounds. We can see the difference with a cylinder, still described as $f(x) = 1$, revolved around the y-axis in Figure 5.5.

For a solid of revolution about the y-axis, we use the shell method. The shell method captures the volume of successive cylinders of varying height, per the describing function, and sums those values for the total volume. Accordingly, we use a different formula to calculate the volume of the solid,

$$V = 2\pi \int_a^b x f(x) \, dx. \tag{5.36}$$

This formula is implemented in Function 5.15, the `shellmethod` function.

This function, like `discmethod`, uses an embedded function which is the object of integration. There is, however, a crucial difference between the functions. The `shellmethod` function is integrated over the multiplication of four

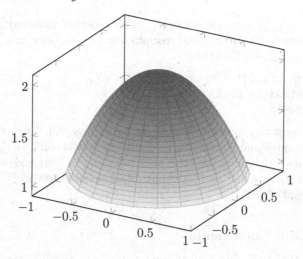

Figure 5.7
Rotated volume around the y-axis

components, π, 2, x, and $f(x)$. The 2 and the π are constants and can be pulled out of the integral, as in the `shellmethod` implementation. This implementation integrates only $xf(x)$. However, the implementation of `discmethod` leaves the constant π inside the function of integration. This increases, unnecessarily, the number of operations in each function evaluation. A more efficient implementation would use underlying calculus to reduce the number of operations necessary.

We can use the volume of the cylinder to again check our `shellmethod` implementation.

```
> f <- function(x) { return(rep(1, length(x))) }
> shellmethod(f, 0, 1)
[1] 3.141593
```

Like with the disc method, the result is π units of volume. However, constants are unique for producing essentially identical volumes when revolved around either the x- or y-axis. In general, we should not expect the same volume from both methods since each represents a different process. For instance, compare the solid of revolution about the x- and y-axes for the function $f(x) = x$.

```
> f <- function(x) { return(x) }
> discmethod(f, 0, 1)
[1] 1.047171
> shellmethod(f, 0, 1)
[1] 2.0945
```

As a final example of the revolved volumes, we have the function described by $f(x) = \sqrt{2-x}$ over the area $[1, 2]$. This function, revolved around the y-axis is shown in Figure 5.7 and resembles a rocket's nose cone. We can find the volume with `shellmethod`:

```
> f <- function(x) { return(sqrt(2 - x)) }
> shellmethod(f, 0, 1)
[1] 3.613796
```

Other geometric measurements rely on the integral. Two of note are the length of a curve and the surface area of the revolved solid. Both are integrals of functions of a known function and both can be implemented in similar manner to the disc and shell methods. Most calculus textbooks provide these formulae and others.

5.5.2   Gini Coefficients

In the last section, our applied example of numerical integration was still fairly abstract, focusing on definitions and idealized shapes. Our second example comes from social science, where we use numerical integration to better understand income inequality, the gap between the haves and the have-nots. The Gini coefficient, sometimes called the Gini index, is a statistical measure, derived from the integral, for that gap.[2]

In economics and sociology, the Lorenz curve for income, given as $L(x)$, describes the portion of the total income that is earned by the lowest-earning x percent of the population. For this curve, over $[0, 1]$, the value of $L(x)$ is the cumulative sum of the amount earned by the portion of the population under x. Accordingly, $L(0) = 0$ and $L(1) = 1$, and the value of $L(x)$ is a percentage, $0 \leq L(x) \leq 1$. Also, Lorenz curves are concave, so for $a \leq b$, then $L(a) \leq L(b)$. Similar measures can be created for wealth, or noneconomic measures, such as amount of education; but, normally Lorenz curves describe income. A sample Lorenz curve where $L(x) = x^2$ is shown in Figure 5.8.

Among a population where there is an exactly equal distribution of income, the Lorenz curve follows the line of perfect equality so that $L(x) = x$. The equal distribution leads to a situation where every subgroup of size x_* percent will have exactly x_* percent of the total income. Graphically, the Lorenz curve is the straight line between 0 and 1.

The Gini coefficient captures how far from equal distribution the actual distribution has fallen. For a given Lorenz curve, $L(x)$, the Gini coefficient is,

$$G_L = 2 \int_0^1 x - L(x) \, dx. \tag{5.37}$$

The Gini coefficient is the area between the straight line of perfect equality and the Lorenz curve. The coefficient is multiplied by 2 to give it a range over

---

[2]For more information on income inequality and the Gini coefficient, Lambert (2001, 27–29) provides a good introduction.

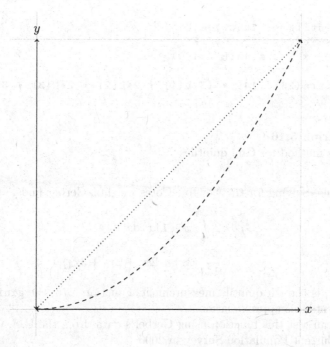

Figure 5.8
Lorenz curve

$[0, 1]$; otherwise, the peak would be $1/2$. Lower Gini coefficient values are more equal population groups and higher values are less equal.

Using a Lorenz curve of x^2, shown in Figure 5.8, we can estimate both the amount of income earned by different percentile groups of the population and the Gini coefficient. For instance, the bottom half of the population has an income of $L(1/2) = (1/2)^2 = .25$, taking in $1/4$ of all the income. The bottom 90 percent receives $L(.9) = (.9)^2 = .81$, leaving the top 10 percent with 19 percent of the income. Overall, the Gini coefficient for this population is

$$G_L = 2 \int_0^1 x - L(x) \, dx = 1/3 \qquad (5.38)$$

Unfortunately, in practice, the actual Lorenz curve is usually unknown. Many statistical agencies produce quintile estimates for the percentage of the population at the 20th, 40th, 60th, and 80th percentiles. We might use the trapezoid or midpoint methods to find the area under this curve. But because the function Lorenz curve is guaranteed to be convex, we are guaranteed to underestimate area of the Gini coefficient using linear functions.

Gerber (2007) presents a method to calculate the Gini coefficient based on published data for income quintiles and using an open four-point Newton–

```
giniquintile <- function(L) {
    x <- c(.2, .4, .6, .8)
    L <- x - cumsum(L / 100)

    return(25 / 144 * (3*L[1] + 2*L[2] + 2*L[3] + 3*L[4]))
}
```

R Function 5.16
Gerber's method for Gini quintiles

Cotes rule. Solving for the Newton–Cotes weights, Gerber finds,

$$G_L = 2 \int_0^1 x - L(x) \, dx \tag{5.39}$$

$$\approx \frac{25}{144} (3x_1 + 2x_2 + 2x_3 + 3x_4), \tag{5.40}$$

where x_i is the ith quintile measurement. Function 5.16, the `giniquintile`, implements this approximation.

We can test this function using Gerber's data, from the U.S. Census Bureau's Current Population Survey in 2000.

```
> L <- c(4.3, 9.8, 15.4, 22.7)
> giniquintile(L)
[1] 0.4223958
```

Using more recent quintile data, the U.S. Census Bureau provides that the quintile measures for 2015 are 3.6, 9.2, 15.1, and 23.2 percent.

```
> L <- c(3.6, 9.2, 15.1, 23.2)
> giniquintile(L)
[1] 0.4418403
```

The quintile rule for Gini coefficients is an excellent demonstration of a custom integration rule to address a specific problem. With this, a relatively good estimate for the Gini coefficient can be had from existing data, with only 14 basic arithmetic operations, and no more complicated function evaluations. Developing specific rules requires time and inclination, but if an integral may be evaluated enough times to justify it, a custom rule can be developed to make the effort worthwhile.

Comments

We have introduced several approaches and many methods for integrating functions numerically in R. Like all other numerical methods, each has some specific advantages and disadvantages and understanding the underlying problem goes a long way toward selecting which approach to use. If we are interested in a stock approach, R includes a fantastic integration function, called `integrate`.

The R `integrate` function is based on an adaptive integration algorithm, internally relying on the QUADPACK family of algorithms, commonly used in numerical computing applications. The function itself is robust enough to handle most numerical integrations and the calling structure is similar to the integration routines we created in this chapter. For instance, assuming $f(x) = \cos^2 x + 1$, we can integrate over the domain from 0 to π using the trapezoid method or the R internals integrator essentially the same way:

```
> f <- function(x) { cos(x)^2 + 1 }
> trap(f, 0, pi)
[1] 4.712389
> integrate(f, 0, pi)
4.712389 with absolute error < 5.2e-14
```

Like the integrators we created in this chapter, `integrate` expects the implementation of $f(x)$ to be vectorized, accepting a vector of x values and returning a vector of y values: The `integrate` function, in addition to returning the estimate, also returns the error estimate. We can use the `str` function to break down the results from the `integrate` function:

```
> str(integrate(cos, 0, pi))
List of 5
 $ value       : num 4.4e-17
 $ abs.error   : num 2.21e-14
 $ subdivisions: int 1
 $ message     : chr "OK"
 $ call        : language integrate(f = cos, lower = 0, upper = pi)
 - attr(*, "class")= chr "integrate"
```

In addition to the results and the error estimate, the `integrate` function also includes a number of other pieces of information about the integration, such as subdivisions count, that represents how many times the integration domain was divided during the integration process. That can be used as informational or even diagnostic if the integrator seems to be misbehaving. In addition, the `integrate` presents a few additional options for fine-tuning the integration process, by fixing the maximum number of subdivisions or the requested error tolerance in the result. The largest drawback to the `integrate` function is that it is limited to one-dimensional integrals and cannot handle any multidimensional integrations.

There is also a package called `gaussquad` for producing Gaussian integrations. The package supports Gauss–Legendre, Gauss–Laguerre, Gauss–Hermite, Gauss–Chebyshev, and numerous other Gaussian integration rulesets. The package separates out the evaluation of the integral and allows for the separate production of Gaussian integration points and weights. However, the format is not compatible with the integration points and weights storage used in this text. Like the `integrate` function, the integrators included with the `gaussquad` package only support integration over a single variable.

If we are interested in integrating multidimensional integrals, there is not a function in base R to handle the problem. However, several add-on packages exist that can perform multidimensional integration. By extending the idea of "quadrature" which attempts to fit a curve with quadrilaterals to three dimensions, we attempt to fit a surface with measurable cubes, a process called "cubature." Accordingly, the first package to note is `cubature`, which uses hypercubes to fit multidimensional integration problems. Second, the `R2Cuba` package also provides a Monte Carlo-based integrator through the `vegas` function and a deterministic integrator through the `cuhre` function, another adaptive multivariate integrator. The `R2Cuba` package provides several other variants, as well.

Beyond the straightforward `integrate` function, there are additional tools for integration available in a number of R packages. These integration tools primarily focus on numerical integrations, since R is designed for numerical computations. Other programs, notably SAGE and Mathematica, are better equipped to handle symbolic computing. Nevertheless, R actually includes a facility for symbolic differentiation. The `deriv` function accepts an expression and can find the first derivative symbolically. However, the `deriv` function is limited to elementary functions, such as polynomials and trigonometric functions. In addition, the results require some work to understand.

```
> deriv(~2 * log(x) - 3 * x^2, "x")
expression({
    .value <- 2 * log(x) - 3 * x^2
    .grad <- array(0, c(length(.value), 1L), list(NULL, c("x")))
    .grad[, "x"] <- 2 * (1/x) - 3 * (2 * x)
    attr(.value, "gradient") <- .grad
    .value
})
```

The `Deriv` package, also available for R, includes a more powerful function for symbolic differentiation. The package supports more complex derivations, beyond the elementary functions, and allows for functions to be specified as expressions, as native R functions, or using R's native formula specification:

```
> library(Deriv)
> Deriv(~2 * log(x) - 3 * x^2, "x")
2/x - 6 * x
```

Despite the availability of these tools, we are probably better off using a dedicated symbolic mathematics program for symbolic computation. SAGE is available for free, like R, and Mathematica is inexpensive for students and educators.

The suite of integration tools presented in this chapter is diverse and each has a place for use. In most cases, the `integrate` function will meet the requirement over one variable. When it breaks down, these alternatives can provide a basis for a revised solution.

Exercises

1. Given a computing environment with a machine epsilon of 0.001, what is the optimal step size for a finite differences method?

2. Describe the relative advantages of the forward, backward, and central differences methods for differentiating functions.

3. Write an equation for a right-hand endpoint rule. Explain the difference between a right-hand endpoint rule and a left-hand endpoint rule. Calculate the maximum error for each. How does this compare to the midpoint rule?

4. Calculate the trapezoid rule for $f(x) = \sin^2 x$ over the domain $x \in [0, \pi]$ for $n = 5$ panels.

5. Write an R function that can calculate a left-hand endpoint rule or a right-hand endpoint rule based on an option.

6. Write an R function that implements a closed Newton–Cotes rule for five interpolating points.

7. Describe the relationship between Newton–Cotes rules and the Vandermonde matrix.

8. The Gauss–Legendre integration given in Section 5.3.1 is only valid over the interval $[-1, 1]$. How can you use this method to find the value of $\int_0^1 x^2 + 1 \, dx$?

9. Implement an algrorithm to automatically implement the method given in problem 8.

10. Rewrite the Romberg integrator as a recursive function.

11. Use a Monte Carlo method to estimate the value of π. Use a different method to estimate the value and describe which method is better and why.

12. Write a Monte Carlo method that returns the estimate and a 95 percent confidence interval for the estimate.

13. Rewrite the Monte Carlo method, `mcint`, to allow for integration over negative regions.

14. Write a Monte Carlo method that integrates over functions of 4 variables.

15. Is it possible to extend the midpoint method to functions over multiple variables?

16. Given the results of 15, implement a function that finds $\int_0^1 \int_0^1 x^2 y^2 \, dx \, dy$.

6

Root Finding and Optimization

6.1 One-Dimensional Root Finding

Root finding is a critical skill we learn in algebra. We learn to find roots by factoring polynomials and solving when $y = 0$. We also learn to guess where a root might be by visually inspecting the graph of a function. For functions that are not polynomials, or well defined like the trigonometric functions, finding a root analytically requires a significant amount of time. If the root represents a real value to a real problem that must be solved repeatedly, finding a numerical solution for the root that is good enough is certainly an acceptable solution.

Root finding and optimization are closely related problems. In both, we seek solutions to functions, linear and nonlinear, that meet some requirement. Accordingly, the solutions for each are similar and so we will begin with the root finding methods. We start with the most elementary, the bisection method, which resembles a children's game. We will also look at the Newton–Raphson method, an iterative approach that uses the derivative to estimate a better result for the next iteration. Finally, we will look at the secant method, which is similar to the Newton–Raphson method, but uses an estimate for the derivative, as well. These methods are designed to find the real roots of an equation. These methods do not find imaginary roots. In addition, each of these methods shares the same flaw of identifying only one root, instead of all roots of a function.

6.1.1 Bisection Method

A popular children's game has one player hide something and another player have to find it based on the being told "hotter" when closer and "colder" when further away. Some children play an equivalent game with numbers where one player picks a number in some range, such as 1 to 100, and another player has to guess the number. The first player will tell the second if the guess is too low or too high. Eventually, the guesser should be able to narrow down the range of options and guess the number exactly. Better players realize earlier the best approach is to halve the range with each turn, which gives the fastest convergence rate.

We can use a similar approach to find the real roots of any continuous

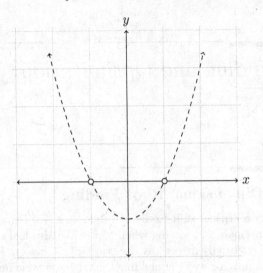

Figure 6.1
The roots of $f(x) = x^2 - 1$

function. The upper and lower bounds for the guess-a-number game creates a reduced domain for the solution. The endpoints are called brackets and we call the process itself bracketing a solution. With any continuous function, we know from Bolzano's theorem that if $f(a)$ exists and is greater than 0 and if $f(b)$ exists and is less than 0, then there exists some value c such that $f(c) = 0$. This is actually a special case of the intermediate value theorem. Of course, there may be more than one value for c meeting the requirements, but we are guaranteed that at least one exists.

The bisection method uses Bolzano's theorem to find a single real root for a given function. Unfortunately for us, it is not intelligent and expects us to give it the initial bracketing parameters. With each iteration through, the distance between the bracket's endpoints is cut in half and the function is evaluated at the new midpoint. If the sign of $f(x)$ evaluated at that midpoint is positive and the left endpoint is positive, we can no longer guarantee a root is in the domain delineated by that endpoint and the midpoint, so we assume it is in the other half and begin the process again. If the signs are different, then we can assume there is a root in the current domain, we discard the other half, and try again. Importantly, we must remember we may be throwing away real roots. The only thing we know is where we can guarantee the existence of a root, not where we cannot.

As we can see from Figure 6.1, showing the graph of $f(x) = x^2 - 1$, we know there are two roots: one is 1 and one is -1. If we start the bisection algorithm with $a = 0.5$ and $b = 1.25$, then $f(a) = -0.75$ and $f(b) = 0.5625$. We have clearly bracketed a real root with a positive value and a negative

```
bisection <- function(f, a, b, tol = 1e-3, m = 100) {
    iter <- 0
    f.a <- f(a)
    f.b <- f(b)

    while (abs(b - a) > tol) {
        iter <- iter + 1
        if (iter > m) {
            warning("iterations maximum exceeded")
            break
        }
        xmid <- (a + b) / 2
        ymid <-  f(xmid)
        if (f.a * ymid > 0) {
            a <- xmid
            f.a <- ymid
        } else {
            b <- xmid
            f.b <- ymid
        }
    }

    ## Interpolate a midpoint for return value
    root <- (a + b) / 2
    return(root)
}
```

R Function 6.1
Finding roots via bisection

value. But if we used $a = -2$ and $b = -2$, we would get $f(a) = f(b) = 3$. We can clearly infer from the graph there are two roots between a and b, but unable to bracket either root, the bisection method would fail with these parameters.

The entire process repeats until the distance between the two endpoints is less than the tolerance specified. Then we know we have found a root, which may very well be at one of these endpoints, but is definitely somewhere in the domain delineated by the final bracketing. The `bisection` function, shown in Function 6.1 implements this algorithm. At the final step, the function finds the midpoint of the domain and returns that value, as the average of all potential values in the range, presumably the best option.

```
> f <- function(x) { x^2 - 1 }
> bisection(f, .5, 1.25, tol = 1e-3)
[1] 0.9998779
> bisection(f, .5, 1.25, tol = 1e-6)
```

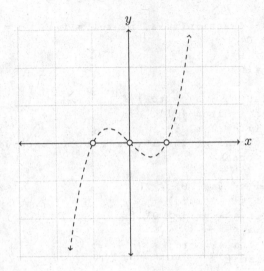

Figure 6.2
The roots of $f(x) = x^3 - x$

```
[1] 0.9999999
> bisection(f, .5, 1.25, tol = 1e-9)
[1] 1
```

For a different example, take $f(x) = x(x-1)(x+1) = x^3 - x$. Analytically, it is clear there are three real roots at -1, 0, and 1. These are shown graphically in Figure 6.2. However, the bisection algorithm will only find one of them. Further, it is found deterministically, so setting the parameters of the algorithm may change which root is found, but upon multiple runs with the same parameters, the same result should be returned.

```
> f <- function(x) { x^3 - x }
> bisection(f, -2, 1.25, tol = 1e-6)
[1] -0.9999997
> bisection(f, -.5, 1.25, tol = 1e-6)
[1] 1.788139e-07
> bisection(f, -2, 1.25, tol = 1e-6)
[1] -0.9999997
```

The same phenomenon can be seen with periodic functions, such as $\sin x$. The sine function has a real root at $k\pi$ where k is any integer. Given a broad enough range, there are a number of roots to find. Due to the continuous splitting of the range, we might assume, given multiple roots, that one close to the middle should be the one found. But this does not always work as the sine example shows:

```
> bisection(sin, 1, 7, tol = 1e-6)
[1] 3.141593
> bisection(sin, -50, 100, tol = 1e-6)
[1] -9.424778
> bisection(sin, -1000, 2000, tol = 1e-6)
[1] 1721.593
```

We can also find a singularity that is not a root. The tangent function is undefined at $\pi/2$. However, just below $\pi/2$, the tangent is positive and just below, the value is negative. This meets the strict requirement of bracketing and if the bisection method were applied with an initial domain of $[1, 2]$, the method will find a root at $\pi/2$:

```
> bisection(tan, 1, 2)
[1] 1.570801
> bisection(tan, -1, 1)
[1] -0.0004882812
```

Of course, it can also find the real root at 0, if the bracket is specified to include it.

Despite the limitations of the bisection algorithm, it is very powerful. The error is explicitly bounded by the tolerance specified. Because of our explicit guarantee provided by Bolzano's theorem, our assurance that there is a root between the endpoints of the domain is secure. As the range gets smaller, our guarantee remains in place. As remarkable as this is, the algorithm's efficiency is even more impressive. The algorithm will continue halving the domain until the domain is narrower than the tolerance. Our implementation includes the safety valve to stop processing if a certain limit had been reached. But if the safety valve does not release, the algorithm will just keep going. The number of times through the loop is,

$$n \geq \log_2 \frac{b-a}{t} - 1, \tag{6.1}$$

where t is the specified error tolerance. Using the number of function evaluations as our metric for efficiency, there are n evaluations, one for each division of the domain, plus 2 for the initial endpoints. For example, any function, if the initial a and b values are separated by 1, a result within 1 one-millionth can be achieved with only 21 function evaluations. If we are willing to accept finding and separately identifying different roots, allowing bisection to narrow them down, the algorithm is a suitable solution.

6.1.2 Newton–Raphson Method

The Newton–Raphson method for finding real roots of a function, $f(x)$ works differently from the bisection method. Provided we also know the first derivative of the function, $f'(x)$, we can refine our estimate of the root based, again, on Taylor's formula. Taylor's formula provides that we can approximate the

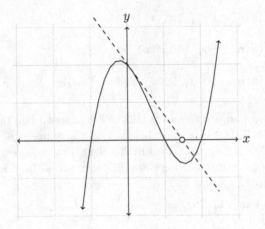

Figure 6.3
Finding a real root of $x^3 - 2x^2 - x + 2$

derivative of a function at x. We used this before to estimate the error in the finite differences method for the derivative in Section 5.1.1. Now, we can use the same process to find roots.

Instead of knowing that we have a real root within some fixed bounds, we start with a single estimate of the real root, x_0, and we know that the slope of the tangent at x_0 is,

$$x_{n+1} = x_n - \frac{f(x_n)}{f'(x_n)}. \tag{6.2}$$

We can iterate over equation 6.2 and receive progressively more accurate results. In fact, we can intuit part of the reasoning, because as $f(x)$ is closer to 0, the step size between x_n and x_{n+1} gets progressively smaller. The first step is illustrated in Figure 6.3.

We can continue iterating until the difference between x_n and x_{n+1} is within some specified tolerance. The `newton` function in Function 6.2 implements this algorithm. We can see this in action with the function $f(x) = x^2 - 1$ and its derivative, $f'(p) = 2x$:

```
> f <- function(x) { x^2 - 1 }
> fp <- function(x) { 2 * x }
> newton(f, fp, 1.25, tol = 1e-3)
[1] 1
> newton(f, fp, -1100, tol = 1e-6)
[1] -1
> newton(f, fp, 1e6, tol = 1e-9)
[1] 1
```

Even with initial starting points both near and quite far away from the true

```
newton <- function(f, fp, x, tol = 1e-3, m = 100) {
    iter <- 0

    oldx <- x
    x <- oldx + 10 * tol

    while(abs(x - oldx) > tol) {
        iter <- iter + 1
        if(iter > m)
            stop("No solution found")
        oldx <- x
        x <- x - f(x) / fp(x)
    }

    return(x)
}
```

R Function 6.2
Finding roots via the Newton–Raphson method

values of 1 and -1, the Newton–Raphson method has converged on correct values.

The advantages of using Newton–Raphson method are that we are not required to bracket a root, as we are with the bisection method. We must only provide an estimate for the root, which need not be particularly good. This points to another advantage over the bisection method. Newton–Raphson can identify a root that only grazes the x-axis, without crossing. For example, let $f(x) = (x-1)(x-.1) = x^2 - 2x + 1$. For this function, there are two roots, both at 1. However, it is impossible to bracket the root between two values of $f(x)$ that are positive and negative since $f(x)$ is never negative.

```
> f <- function(x) { x^2 - 2 * x + 1 }
> fp <- function(x) { 2 * x - 2 }
> newton(f, fp, 1.25, tol = 1e-3)
[1] 1.000508
> newton(f, fp, -1100, tol = 1e-6)
[1] 0.9999995
> newton(f, fp, 1e6, tol = 1e-9)
[1] 1
```

The Newton–Raphson method is also fast, converging quadratically to a solution. But the Newton–Raphson method also presents some downsides. First among them, we must identify, analytically, the derivative the function we are finding the roots of, which may not be feasible.

In addition, the Newton–Raphson method, when not converging rapidly can misbehave, even in elementary cases. For instance, consider the example of

$f(x) = x^2 - 1$, again. We know there are roots at $x = 1$ and $x = -1$, and both were found by the `newton` function with widely different starting points. But if the starting point were $x_i = 1$, the process will break down immediately because $f'(1) = 0$ and the Newton–Raphson method must divide by $f'(x)$ with each iteration.

```
> newton(f, fp, 0, tol = 1e-3)
[1] 0.9990332
```

If the Newton–Raphson method should stumble upon any other point where the derivative is 0, the process will break down there, too. In addition, if $f'(x)$ is very close to 0, the Newton–Raphson method can diverge catastrophically. This is because the denominator of $f(x)/f'(x)$ is very small causing the overall term to grow very quickly.

Because of this potential divergence, our implementation "fails gracefully" if our stopping requirements are not met. With the bisection method, the range found at any point is sure to have a root within it. Specifying either the tolerance or the maximum number of iterations defines the maximum error of the result. Our implementation offers setting both as a convenience to the user. But the maximum iteration count is used as the safety valve on the process, simply to provide an exit if the tolerance seems unreachable. The Newton–Raphson method is not guaranteed to be anywhere near a root when the safety valve opens. Accordingly, the function gives an error and fails to return a usable value. This is better than returning the current iterate, since we cannot be sure its an even mediocre solution.

We can also test the Newton–Raphson method on the sine function, like we did the bisection method:

```
> newton(sin, cos, 2, tol = 1e-6)
[1] 3.141593
> newton(sin, cos, pi, tol = 1e-6)
[1] 3.141593
> newton(sin, cos, pi/2, tol = 1e-6)
[1] 99978.04
```

This example is interesting, because we again see how a root finding method can have difficulty with a periodic function. We typically set the value of x_0 to something near where we believe a root to be. This continues our practice of ensuring we understand the function we are operating on and how our numerical methods may work with them.

In the last sine example, we selected an x_0 value of $\pi/2$. This is a good guess, since there is a real root at 0 and π, both of which are only half a π away. In reality, the derivative function $\cos(\pi/2) = 0$, so this should break down. But in practice, the machine misses and produces a nonzero value:

```
> cos(pi/2)
[1] 6.123234e-17
```

Dividing by that produces a larger error in the result. Eventually, the Newton–

```
secant <- function(f, x, tol = 1e-3, m = 100) {
    i <- 0

    oldx <- x
    oldfx <- f(x)
    x <- oldx + 10 * tol

    while(abs(x - oldx) > tol) {
        i <- i + 1
        if (i > m)
            stop("No solution found")

        fx <- f(x)
        newx <- x - fx * ((x - oldx) / (fx - oldfx))
        oldx <- x
        oldfx <- fx
        x <- newx
    }

    return(x)
}
```

R Function 6.3
Finding roots via the secant method

Raphson method settles down and gives the result above. However, the result is far from what we expected.

In other cases, the iterative process may diverge away from any value. While the sine function eventually settled, others may not. Some functions may continue off and never return a result. Others could end up in a loop if there is a value x_i that repeats itself. Obviously, if the value appears twice in the sequence, whatever came after it will come again and the process will never converge.

6.1.3   Secant Method

The hardest part of working with the Newton–Raphson method is the need to find the first derivative of the function of which we are root finding. In Section 5.1.1, we found a method that allows us to approximate the first derivative using a finite differences formula. We might ask if we could apply the same process as part of the Newton–Raphson formula to find the result without first analytically deriving the first derivative. We would be right to ask such a question and the secant method implements this change.

The secant method is given in Function 6.3, the `secant` function. This implementation does not call the `findiff` function, but does use the same ideas and essentially reimplements the finite differences formula. For this implementation, that design decision eliminates some otherwise unnecessary function evaluations.

One other substantial difference here is the selection of the step-size for finite differences. Initially. the step size is set to 10 times the error tolerance. From then on, the step size is set dynamically based on the distance between iterations. That is, the step size changes based on how the function behaves around the root in play. In one sense, this is an adaptive algorithm that responds to the function it is operating on.

Like both the bisection and the Newton–Raphson methods, we begin by applying the secant method to the function $f(x) = x^2 - 1$:

```
> f <- function(x) { x^2 - 1 }
> secant(f, 1.25, tol = 1e-3)
[1] 1
> secant(f, -1100, tol = 1e-6)
[1] -1
> secant(f, 1e6, tol = 1e-9)
[1] 1
```

The secant method finds the same results as the Newton–Raphson method when x_0 is identical. However, this will not be true in the general case. We can also test the secant method on the sine function, like we have our other root finding methods:

```
> secant(sin, 2, tol = 1e-6)
[1] 3.141593
> secant(sin, pi, tol = 1e-6)
[1] 3.141593
> secant(sin, pi/2, tol = 1e-6)
[1] -44627819
```

In general, the secant method offers some benefits over other methods. First, like the Newton–Raphson method and unlike the bisection method, it is not necessary to bracket a root to find the value. Unlike the Newton–Raphson method, it is not necessary to analytically find the first derivative, which can be challenging and cannot be automated in all cases.

But the secant method does have some drawbacks. It will diverge or not find the root in the same cases as the Newton–Raphson method. Further, while it is fast, the convergence rate is less than quadratic, leaving it slower than the Newton–Raphson method. However, in most cases, if the initial root selection is close enough to the real root of interest, the secant method will probably converge to the correct value after a suitable number of iterations.

6.2 Minimization and Maximization

True optimization, unlike root finding, leaves us looking for a value of x where y is also unknown. We are just looking for wherever the y value is smallest or largest, usually within a region. In this section, we will look at a handful of methods for finding the minima and maxima, or the extrema, of functions of single variables. These generally focus on local extrema.

Because of the close relationship to root finding, we will observe that the methods used for finding extrema are conceptually similar to root finding. In fact, when a continuous differentiable function reaches a minimum or a maximum, its first derivative will be 0. Finding the roots of the first derivative is one approach to optimization. However, methods that can address sharp bends and corners in the underlying function are appreciated and we want to use this conceptual similarity.

We extend this conceptual similarity with a set of parallel examples. Initially, we will review a minimization technique that will find a minimum if it is bracketed, like the bisection method for root finding. Then we will move to gradient descent, a minimization technique that relies on knowing the first derivative of the function in question, like the Newton–Raphson method. Knowing the first derivative, if possible, informs the minimization, as the zeros of the derivative are the local minima and maxima of the function.

6.2.1 Golden Section Search

Minimization and maximization problems are usually handled the same way. However, wherever a minimizer is looking for a smaller solution, a maximizer is looking for a larger solution. For the golden section search, we will implement a maximizer and note that a golden section search for minimizer will be very nearly identical. As noted, the methods for extrema searching in functions of one variable are very similar in both spirit and process to root finding.

With the golden section search, we must define some constraints. First, we start with a function $f(x)$. We also identify bounds for the local maximum value of a function. Through other means, typically a visual inspection, we can identify a point that seems to be a local maximum, x', such that

$$f(x') \geq f(x), \ x \in [a, b]. \tag{6.3}$$

We can also probably guess at the values for a and b, which need not be the broadest selection, but must satisfy equation 6.3.

These points define a domain for $f(x)$ over which the function must be *unimodal*, that is, the function must be strictly increasing, reach its local maximum, then be strictly decreasing. Over the domain, the function will generally behave like a parabola, rather than producing oscillating function values. This is a relatively strong constraint.

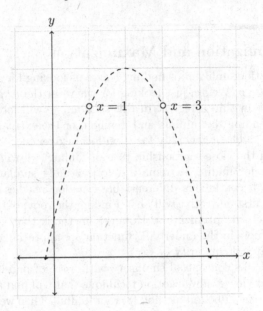

Figure 6.4
Golden section search on $f(x) = -x^2 + 4x + 1$

For example, we will consider the example of $f(x) = -x^2 + 4x + 1$, shown in Figure 6.4. The function is a parabola and the maximum can be found analytically. The maximum is clearly visible somewhere between $a = 1$ and $b = 3$, and this is a suitable bound for bracketing the maximum.

Once a domain within which the maximum is contained is identified, we can select some third point where $f(x)$ is definitely greater than the $f(a)$ and $f(b)$ due to the unimodality requirement. We call this point is $a^\star$. Then, we can select one more point, $b^\star \in [a^\star, b]$ also in the domain. For the same reason, we know that $f(b^\star)$ is greater than both $f(a)$ and $f(b)$, but there is no predefined relationship between $f(a^\star)$ and $f(b^\star)$.

If $f(a^\star) > f(b^\star)$, then we can replace b with $b^\star$ and the original constraints still hold. In the alternative, if $f(a^\star) < f(b^\star)$, then we can replace a with $a^\star$ and the original constraints still hold. Like with the bisection method, we can repeat the process until the bounds of the maximum, $[a, b]$, are within an acceptably small tolerance.

We have a basic method now, but we still do not have an approach for selecting $a^\star$ and $b^\star$. Instinctively, we might set $a^\star$ initially to be the half-way point between a and b, with $b^\star$ half-way between $a^\star$ and b, continuing the analogy with the bisection search. That would work, but for the fastest convergence, we should mimic the idea and not the process.

First the interval between $a^\star$ and $b^\star$ is guaranteed to be in the new search

```
goldsectmax <- function(f, a, b, tol = 1e-3, m = 100) {
    iter <- 0
    phi <- (sqrt(5) - 1) / 2

    a.star <- b - phi * abs(b - a)
    b.star <- a + phi * abs(b - a)

    while (abs(b - a) > tol) {
        iter <- iter + 1
        if (iter > m) {
            warning("iterations maximum exceeded")
            break
        }

        if(f(a.star) > f(b.star)) {
            b <- b.star
            b.star <- a.star
            a.star <- b - phi * abs(b - a)
        } else {
            a <- a.star
            a.star <- b.star
            b.star <- a + phi * abs(b - a)
        }
    }

    return((a + b) / 2)
}
```

R Function 6.4
Golden section search for local maxima

interval. Second, either the interval between a and a^* or the interval between b and b^* will be in the new search interval. Accordingly, we want to balance those two intervals against each other. To ensure that balance remains fixed, the proportion of $(b - a^*)/(a^* - a)$ should be the golden ratio, or $\phi \approx 1.618$. Further, $(a^* - a)/(b^* - a^*)$ will also be ϕ. Function 6.4, the goldsectmax, implements this function.

We can see the results with the example function, $f(x) = -x^2 + 4x + 1$.

```
> f <- function(x) { -x^2 + 4 * x + 1 }
> goldsectmax(f, 1, 3)
[1] 2.00028
```

Since we know analytically the parabola's maximum value is at $x = 2$, this is a reasonable first blush estimate. The default tolerance of $1/1000$ is sufficiently narrow that we receive a reasonable estimate, without requiring too many trips through the loop. In the case of a parabola, we estimate the maximum

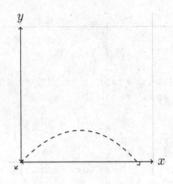

Figure 6.5
Golden section search on $f(x) = \sin x - x^2$

by treating the curve like a parabola. The golden section search works just as well with more complex curves. For instance, we can set $f(x) = \sin x - x^2$, shown in Figure 6.5.

This function, which is not a parabola, has a peak value somewhere between 0 and 1, so we use those as the starting points for the golden section search.

```
> f <- function(x) { sin(x) - x^2 }
> goldsectmax(f, 0, 1)
[1] 0.45003
```

The golden section search is a good algorithm for maximizing functions of one variable, if the maximum can be bounded. Further, the logic to reverse the search, to find a minimum value, is straightforward. For finding maxima without bounding the search area, however, different methods are used.

6.2.2   Gradient Descent

Gradient descent is the next optimization algorithm we are going to look at. The algorithm begins with Rolle's theorem, which states that if $f(a) = f(b)$, and a and b are not equal, there is some value of x, which we will call c, such that $f'(c) = 0$. More concretely, if function goes up and then down, or down and then up, there is some point in between going up and going down where the function changes direction and at those peaks and troughs, the first derivative of the function is 0. Like most of calculus, this also requires that $f(x)$ be both continuous and differentiable.

If we know the first derivative, we could, analytically or numerically, solve this for the roots. The values found would be local extrema for the function. But this approach does not extend itself very well. A different approach that is similar to the Newton–Raphson method for root finding requires only know-

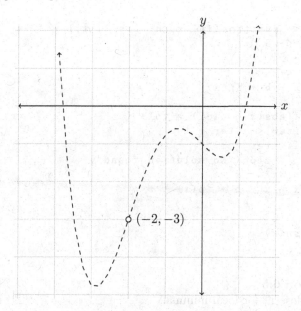

Figure 6.6
Gradient descent on $\frac{1}{4}x^4 + x^3 - x - 1$

ing the first derivative, which must be found analytically. Like the Newton–Raphson method, gradient descent is an iterative approach, that requires an initial value, x, to start, and also requires a value h, which is a step size used to control the iterations. Like many other iterative methods, parameters to provide·for the maximum number of iterations and the error tolerance are included.

Presuming the local region is not constant-valued, the gradient descent method uses the derivative at x and takes a step down, of size h, in the direction of the slope. After the step, the process is repeated using the new point as x. This process continues, but as the function slides down a slope, if the function begins to reach a trough, the derivative will start shrinking resulting in smaller changes in x. As the change in x decreases below the tolerance value, we can be assured we have reached a local minimum.

Of course, the process can be reversed to take a step up to find local maxima. Then the process is called gradient ascent. The gradient descent process is implemented in the function **graddsc**, given in Function 6.5, along with the gradient ascent function, **gradasc**, shown in Function 6.6. These two functions are identical except for their names and the direction of the shift, given by the signs on lines 5 and 12. In general, statements about gradient

```
graddsc <- function(fp, x, h = 1e-3, tol = 1e-4, m = 1e3) {
    iter <- 0

    oldx <- x
    x = x - h * fp(x)

    while(abs(x - oldx) > tol) {
        iter <- iter + 1
        if(iter > m)
            stop("No solution found")
        oldx <- x
        x = x - h * fp(x)
    }

    return(x)
}
```

R Function 6.5
Gradient descent for local minima

descent will apply to gradient ascent, with the note that sign or direction may change as appropriate.[1]

We will begin by showing a quartic function. A quartic function will go up and down several times. For our example, the more the better, therefore,

$$\frac{1}{4}x^4 + x^3 - x - 1. \tag{6.4}$$

This function is shown in Figure 6.6. We can see from the graph that there are two local minima and one local, finite, maximum. The first derivative of equation 6.4 is $x^3 + 3x^2 - 1$. We will use this function to test gradient descent. And for illustration, we will begin with $x = -2$, as is highlighted in the figure.

```
> fp <- function(x) { x^3 + 3*x^2 - 1 }
> graddsc(fp, x = -2, h = 1/100)
[1] -2.878225
```

Simply switching which function, and therefore the sign of the change in x, results in the local maxima being found.

```
> gradasc(fp, x = -2, h = 1/100)
[1] -0.6563222
```

Gradient descent will find a local minimum if it exists, and will be accurate within the tolerance specified. However, there are some cases where the function may not end in the correct result. First, when searching for a

---

[1]In addition, changing a minimization function a maximization function is usually trivial, but the maximizing function here will serve special use.

```
gradasc <- function(fp, x, h = 1e-3, tol = 1e-4, m = 1e3) {
    iter <- 0

    oldx <- x
    x = x + h * fp(x)

    while(abs(x - oldx) > tol) {
        iter <- iter + 1
        if(iter > m)
            stop("No solution found")
        oldx <- x
        x = x + h * fp(x)
    }

    return(x)
}
```

R Function 6.6
Gradient ascent for local maxima

local minimum, if the initial point, x_0, given is a local maximum, then the derivative there is 0. If the derivative is 0, regardless of the given step size, the actual change in x is always 0. Also, if down the slope heads downward into infinity, the gradient descent method will continue down the slope until the safety valve releases. Finally, like the Newton–Raphson method, gradient descent relies on finding the first derivative analytically.

We can see the behavior of gradient descent failing if we give an initial x value of 1. With gradient descent, the function finds the local minimum near 1. However, if gradient ascent is used, the method eventually stops processing as the maximum number of iterations is reached. In the mean time, the method has ascended the slope as x grows, never to return.

```
> graddsc(fp, x = 1, h = 1/100)
[1] 0.5343634
> status <- try(gradasc(fp, x = 1, h = 1/100, m = 100))
> cat(status[1])
Error in while (abs(x - oldx) > tol) { :
  missing value where TRUE/FALSE needed
```

Because of the **stop** command, the output cannot be reproduced as normal and we use the **try** function to capture the status and display it. Despite these problems, gradient descent and gradient ascent functions are excellent for finding local extrema, if the derivative of the subject function is already known.

6.3   Multidimensional Optimization

Finding the minimum or the maximum of a function of a single variable can
be useful, but methods we can extend to more than one dimension will find
more use. Solutions to minimization problems are important in physics, where
celestial mechanics, the study of orbits in space, are solved by solving a minimization. Maximization also solves a number of problems in economics, especially around individual behavior. These are just two examples of optimization
problems, that have consumed a great deal of research in the field of numerical
analysis. Because of this, there are many more methods than we can document
here.

In other areas of numerical analysis, we have found multidimensional extensions of solutions to single variable problems. Some of these are very suc-
cessful and others are just the first step in more than one dimension. However,
some methods cannot be extended. The golden section search does not extend
to more than one dimension because there is no multidimensional analog to
dividing the space, whether by the golden ratio, in halves, or any other way.
However, gradient descent can be extended.

6.3.1   Multidimensional Gradient Descent

Gradient descent can be extended to support functions of more than one
variable by making two basic changes to Function 6.5. First, we must tacitly
allow for the function to accept more than one variable, and by extension,
we must also allow our initialization of the routine to include more than one
variable. This can be done effectively by ensuring the function derivative, the
variable fp, accepts a vector of variable values.

Second, we must evaluate whether the error tolerance exceeds the error
estimate at x. For gradient descent in one variable, we simply check $|x_i - x_{i-1}|$.
If the value were below the tolerance specified, we judged we had succeeded.
To extend that to two or more dimensions, we want to use the vector **x**'s
norm. In Section 3.4, we developed a function vecnorm, that would calculate
the norm of a vector for use in iterative methods over matrices. Now, we will
use vecnorm again to estimate the error of an iterative solution.

Once these two changes are made, we have a multidimensional gradient
descent. This function is given as Function 6.7, the gd function. There are
two distinct differences between gd and graddsc. First, in the while loop, the
condition is changed that the norm of **x** must be below the tolerance. This is
the straightforward extension. Second, the function, after reaching a maximum
number of iterations, returns the current result. The single dimension version
of graddsc would fail gracefully and with an error. Neither is necessarily
correct.

```
gd <- function(fp, x, h = 1e2, tol = 1e-4, m = 1e3) {
    iter <- 0

    oldx <- x
    x = x - h * fp(x)

    while(vecnorm(x - oldx) > tol) {
        iter <- iter + 1
        if(iter > m)
            return(x)
        oldx <- x
        x = x - h * fp(x)
    }

    return(x)
}
```

R Function 6.7
Gradient descent for n dimensions

To show this in action, we will begin with the function,

$$f(x_1, x_2) = (x_1 - 1)^2 + (2x_2 - 2)^2. \qquad (6.5)$$

Shown in Figure 6.7, there is a bowl-shape that has a lowest point at $(1, 1)$. In addition, viewed across the x-axis, the graph is a parabola with center at 1. Viewed across the y-axis, the graph is a different parabola with a center at 1. We can see from equation 6.5, the real minimum seems to be at $(1, 1)$. To solve this, we must find the partial differential equations for each variable input of $f(x_1, \ldots)$ with respect to y. Therefore,

$$\frac{\partial f}{\partial x_1} = 2x_1 - 2 \qquad (6.6)$$

$$\frac{\partial f}{\partial x_2} = 8x_2 - 8. \qquad (6.7)$$

These two equations are given in the function fp.

```
> fp <-function(x) {
+     x1 <- 2 * x[1] - 2
+     x2 <- 8 * x[2] - 8
+
+     return(c(x1, x2))
+ }
```

As we can see, fp returns a two-element vector, one for each variable of $f(x_1, \ldots)$. The gradient descent function can minimize a function given a starting value, step size, and error tolerance. Like all of our iterative functions,

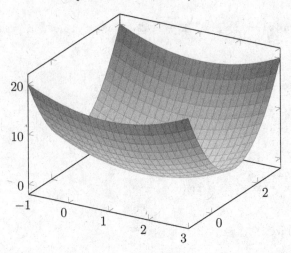

Figure 6.7
Plot of $f(x_1, x_2) = (x_1 - 1)^2 + (2x_2 - 2)^2$

we include a maximum step count to prevent the function from running away.

```
> gd(fp, c(0, 0), 0.05, m = 5)
[1] 0.468559 0.953344
> gd(fp, c(0, 0), 0.05, m = 10)
[1] 0.6861894 0.9963720
> gd(fp, c(0, 0), 0.05, m = 20)
[1] 0.8905810 0.9999781
```

The `gd` comment accepts the derivative function of $f(x_1, \dots)$ and slowly marches it to its lowest point. We can see this in action by setting a very small maximum step count and examining the results. We can also note that both variables are, though at different rates, converging to their respective values, both 1. If we up the step limit, we can get a better result.

```
> gd(fp, c(0, 0), 0.05, m = 1000)
[1] 0.9991405 1.0000000
```

Also, changing the step size can affect convergence. Larger step sizes lead to faster convergence.

```
> gd(fp, c(0, 0), 0.001, m = 100)
[1] 0.1830703 0.5556974
> gd(fp, c(0, 0), 0.01, m = 100)
[1] 0.8700328 0.9997799
> gd(fp, c(0, 0), 0.10, m = 100)
[1] 0.9996755 1.0000000
```

However, if the step size is too large, the process can lead to ridiculous results.

```
> gd(fp, c(0, 0), 0.50, m = 100)
[1] 1.000000e+00 1.546133e+48
```

As is true for other functions relying on the first derivative for calculations, it may not be possible, or at least easy enough, to calculate. This can be an impediment to effective use in multiple dimensions and this leads us to algorithms that can operate on the objective function itself.

6.3.2 Hill Climbing

Wanting to operate directly from the objective function leads us to hill climbing. This function is advantageous if the first derivative of the underlying function is not available or cannot be reasonably calculated. Hill climbing, despite its name, can be used to find either a maximum or a minimum. In addition, there is frequent confusion between the hill climbing and gradient descent. While they are somewhat similar, they are two different algorithms with two different methods of operation internally, and different stopping conditions.

Continuing our study of minimizing functions, hill climbing works by taking an initial condition and evaluating a single neighbor to see if it presents a lower value of the function. If the neighbor is a lower value of the function, the current is updated and a new neighborhood search is started. The process continues until a fixed number of steps are taken since there is no clear ending condition.

Another critical distinction between hill climbing and gradient descent focuses on how the next step is taken. With gradient descent, the algorithm will slide downslope at whatever angle is created by the gradient. When hill climbing, the algorithm will just change one element of the inputs, one of the input parameters, and evaluate it as the neighbor. For a function of a single variable, this distinction is irrelevant. For a function of more than one variable, hill climbing makes all jumps at right angles to each other. Hill climbing is given in Function 6.8, the `hillclimbing` function.

With hill climbing, the decision on which neighborhood point to evaluate is left to the implementor. Different selection processes lead to different variants of the algorithm. This implementation selects an input variable at random, x_i. The change in that variable then is drawn randomly from a uniform distribution with mean of x_i and standard deviation of h, the implied step size. This is a form of stochastic hill climbing, which are hill climbing implementations where the direction is selected at random. Another option for direction selection is to change each input variable and evaluate to see which is closest and jumping to it.

To show hill climbing in action, we will use the Himmelblau function (Himmelblau 1972), which has four minima at approximately $(3, 2)$, $(-2.8, 3.1)$,

```
hillclimbing <- function(f, x, h = 1, m = 1e3) {
    n <- length(x)

    xcurr <- x
    ycurr <- f(x)

    for(i in 1:m) {
        xnext <- xcurr
        i <- ceiling(runif(1, 0, n))
        xnext[i] <- rnorm(1, xcurr[i], h)
        ynext <- f(xnext)
        if(ynext < ycurr) {
            xcurr <- xnext
            ycurr <- ynext
        }
    }

    return(xcurr)
}
```

R Function 6.8
Hill climbing for n dimensions

$(-3.8, -3.3)$, and $(3.6, -1.8)$. The function itself is,

$$f(x) = (x_1^2 + x_2 - 11)^2 + (x_1 + x_2^2 - 7)^2. \tag{6.8}$$

```
> f <- function(x) {
+     (x[1]^2 + x[2] - 11)^2 + (x[1] + x[2]^2 - 7)^2
+ }
> hillclimbing(f, c(0,0))
[1] 2.998820 2.002761
> hillclimbing(f, c(-1,-1))
[1] -2.805738  3.132056
> hillclimbing(f, c(10,10))
[1] -2.802708  3.132264
```

Because the process is stochastic, the results may not necessarily be the same between consecutive runs, even with the same starting parameters.

```
> hillclimbing(f, c(0,0))
[1] -2.805445  3.131193
> hillclimbing(f, c(-1,-1))
[1] -3.781078 -3.290036
> hillclimbing(f, c(10,10))
[1] 2.999725 2.003107
```

One interesting note in this algorithm is the default step size is 1. In the case of hill climbing, the step size parameter does not specify the step size, but rather the standard deviation of the step size taken at each iteration. This can allow for both exceptionally large and exceptionally small steps, especially as the distribution is normal. In practice, the step size parameter should be set for each use based on the objective function.

6.3.3   Simulated Annealing

Gradient descent and hill climbing do an excellent job of finding a nearby minimum. However, both are constrained by their initial condition. These functions will start from the initial condition and steadily move down the slope it is on until it finds a low point. This is excellent for finding a local minimum, but there could be deeper minima, even nearby, that are completely ignored in this process. This suggests we need different algorithms to hunt for global minima. For instance, an algorithm could find a local minimum and, in a sense, look around and see if it can go lower by restarting the process. This could, conceivably, produce global minima, and is the premise behind simulated annealing.

Simulated annealing, as its name suggests, draws upon the metallurgical process of annealing. When annealing, the metal is heated then allowed to cool under very controlled circumstances. This process forces the metal to re-crystallize as it cools, eliminating flaws and strengthening the metal. A similar process happens in your kitchen when chocolate is tempered.

Simulated annealing mimics this process using an initial state, a function to minimize, a temperature analog, and a cooling rate for the temperature analog. The process starts by storing the best estimate, x^*, it has found for a global minimum, which is also the initial state. Then the process looks in the immediate neighborhood for another potential minimum state, x_i. If $f(x^*) > f(x_i)$, then a new best state is recorded. Otherwise, the process tries again.

Generally, this process looks like a simple peek around the neighborhood with the occasional move toward a better position, just like in hill climbing. However, simulated annealing makes a key change. Occasionally, based on a random draw, the function will make a large jump to a position that is worse to see if that position is a better starting point for a downhill run. All the while, the historical best is kept just in case the local minimum found this way is not a better overall option.

The decision to jump to a new point is based on the temperature analog. Internally, simulated anneal maintains the current temperature and reduces it by a small amount for each time through the main loop. As temperature decreases, so should the likelihood of making a large jump. An R function, `sa`, for this is given in Function 6.9. The implementation here is an extension of the hill climbing method given in `hillclimbing`.

In general, simulated annealing leaves several key design decisions in the

```r
sa <- function(f, x, temp = 1e4, rate = 1e-4) {
    step = 1 - rate
    n <- length(x)

    xbest <- xcurr <- xnext <- x
    ybest <- ycurr <- ynext <- f(x)

    while(temp > 1) {
        temp <- temp * step
        i <- ceiling(runif(1, 0, n))
        xnext[i] <- rnorm(1, xcurr[i], temp)
        ynext <- f(xnext)
        accept <- exp(-(ynext - ycurr) / temp)
        if(ynext < ycurr || runif(1) < accept) {
            xcurr <- xnext
            ycurr <- ynext
        }
        if(ynext < ybest) {
            xbest <- xcurr
            ybest <- ycurr
        }
    }

    return(xbest)
}
```

R Function 6.9
Simulated annealing for n dimensions

hands of implementors. The first of these is how to generate a neighboring solution to test. While this implementation of simulated annealing can operate on functions of a single variable, it also operates on functions that accept multivariate vectors as inputs. Therefore, this implementation selects one parameter of the multivariate function, at random using `runif`, then alters that parameter to a random variable. The random variable is selected from a normal distribution centered on the current value of that parameter and with a standard deviation set to the current temperature analog. Other standard deviation selections are feasible, but this option encourages the simulated annealing process to explore the parameter space with large jumps early in the process. The decreasing temperature value as the process progresses discourages large jumps later. A related decision is how to set the cooling rate. This, like the step size of other iterative processes, can greatly affect convergence and the rate of convergence.

A second major design decision is how to define the acceptance probability for jumping to a worse solution. A common approach is to use,

$$a(y_i, y_{i+1}, T) = \exp \frac{y_{i+1} - y_i}{T}, \qquad (6.9)$$

where y_i is the current value of $f(x)$, y_{i+1} is the proposed value, and T is the current temperature. This acceptance probability is used and compared to a single random draw, over $[0, 1]$, for each time through the loop. Higher values of T, at higher temperatures, lead to larger values of the acceptance probability. Therefore, at higher temperatures, jumps are both more likely and more likely to be larger.

We can see this process in action with a univariate polynomial, $f(x) = x^6 - 4x^5 - 7x^4 + 22x^3 + 24x^2 + 2$.

```
> f <- function(x) {
+     x^6 - 4 * x^5 - 7 * x^4 + 22 * x^3 + 24 * x^2 + 2
+ }
> sa(f, 0)
[1] 3.612628
> sa(f, 0)
[1] 3.612546
> sa(f, 0)
[1] 3.612745
```

With the same parameters, the results are nearly the same each time, though are not exact.

Simulated annealing offers some important advantages in optimization. First, it is not necessary to calculate the derivative of the objective function. The example functions provided here are easy to differentiate, but real-world functions are not so generous. A second advantage is that we know, *a priori*, exactly how long the process will take since it is a function of the initial temperature and the cooling rate.

The downsides to simulated annealing, are important. While we know how long the process will take, we also know it will take a long time to produce usable results. Starting with high temperatures and low cooling rates is necessary, especially as the number of variables of $f(x)$ increases. Additionally, we can never really know if the result found is truly the global minimum, or if it is a conveniently found local minimum. However, graphing the function and hand analysis should give some idea where the global minimum is and that information can be used to set the initial starting point for annealing.

Simulated annealing, here, was applied to a continuous function, but it is often applied to discrete functions, as well. In particular, simulated annealing is a favorite approach for solving the traveling salesperson problem, where finding the shortest path encompassing a set of discrete points is the objective. Application to continuous functions, can be hit or miss.[2] Consider, for

---

[2]For an overview of the practice, see Henderson, Jacobson, and Johnson (2003).

example, applying simulated annealing to Ackley's function,

$$f(x) = -20 \exp \frac{1}{5} \sqrt{\frac{1}{2}(x_1^2 + x_2^2)}. \tag{6.10}$$

```
> ackley <- function(x) {
+     -20 * exp(-0.2 * sqrt((x[1]^2 + x[2]^2) / 2))
+ }
> sa(ackley, c(1, 1))
[1] 1 1
> sa(ackley, c(2, -1))
[1]   2 -1
> sa(ackley, c(-1/2, -1/3))
[1] -0.5000000 -0.3333333
```

The results of simulated annealing, in this case, are highly sensitive to the initial conditions and not even usable. We will explore simulated annealing for discrete problems in Section 6.4.2.

6.4   Applications

Optimization has applications across the entire mathematical spectrum. Optimization is used in packing ships for transit, in finding economic equilibria, and in medical imaging. Optimization is such a rich field of research, entire texts have been written solely on optimization. To expand our horizons within a numerical analysis framework, we are going to look at two applications.

For R users well versed in statistics and data science, gradient descent has the ability to estimate generalized linear models. We will show the elegant least squares and gradient descent to show the method. By changing the function optimized, other linear models can be estimated, such as logistic regression or more exotic models. Finally, we will show a different implementation of simulated annealing for solving discrete models. We will implement a solution to the traveling salesperson problem, which can be extended to other computer science problems.

6.4.1   Least Squares

Gradient descent can be used to solve the least squares problem and the method has become popular in machine learning circles. In machine learning, a subfield of computer science, we use a computer for pattern matching and for making predictions. This predictive ability is used by data scientists to create recommendation systems, such as "suggested products" on online retail websites, or in handwriting recognition with the latest portable electronic devices. Machine learning is also a cornerstone of artificial intelligence research, where computer scientists try to create machines that think.

For our purposes, solving the least squares problem will be a different kind of gradient descent. First, there is a change in terminology. What we have traditionally called "step size" is instead called the "learning rate," and it uses the variable `alpha` or α to represent the value. In addition, the function we are attempting to minimize is called the "cost function," typically represented as $J(\theta_1, \theta_2, \ldots)$. In these equations, *theta*$_i$ is the ith element of the input variables. In gradient descent, we called these variables x_i. These changes in terminology allow us to generalize the approach for other types of machine learning, other than least squares approximation.

The cost function is the function that defines how far each observation is from the least squares line. Therefore, the function is,

$$J(\theta_1, \theta_2, \ldots) = \frac{1}{2m} \sum_{i=1}^{m} (\hat{y}_i - y_i)^2, \tag{6.11}$$

where m is the number of observations and $\hat{y}_i$ is the ith estimate on the least squares line. Minimizing the cost function is a matter of applying gradient descent, but gradient descent requires the partial derivatives of y with respect to each θ. For least squares, the partial derivative is,

$$J'(\theta) = \frac{1}{m} X^\mathsf{T} (X\theta - y). \tag{6.12}$$

which yields a vector of gradients we can apply, with an appropriate learning rate, to an initial estimate of the minimum. This method is shown in Function 6.10, the `gdls` function.

It is worth noting the cost function itself is not included. Formal implementations include an evaluation of the cost function as this is the mean squared error, one of the key methods to evaluate the accuracy of a regression.

The `gdls` function requires a value of A and b. It also accepts a learning rate, error tolerance, and maximum number of steps. We can see this in action with the classic iris dataset (Fisher 1936). The iris dataset has become a staple of machine learning examples and is often used to test or demonstrate multivariate classifiers. There are five fields, the iris species, a factor, and measurements for the sepal's length and width and petal's length and width. For this demonstration, we will predict the sepal's length from the width and petal length and width. We begin by creating variables b and A.

```
> head(b <- iris$Sepal.Length)
[1] 5.1 4.9 4.7 4.6 5.0 5.4
> head(A <- matrix(cbind(1, iris$Sepal.Width,
+             iris$Petal.Length, iris$Petal.Width), ncol = 4))
     [,1] [,2] [,3] [,4]
[1,]    1  3.5  1.4  0.2
```

```
gdls <- function(A, b, alpha = 0.05, tol = 1e-6, m = 1e5) {
    iter <- 0
    n <- ncol(A)
    theta <- matrix(rep(0, n))
    oldtheta = theta + 10 * tol

    while(vecnorm(oldtheta - theta) > tol) {
        if((iter <- iter + 1) > m) {
            warning("iterations maximum exceeded")
            return(theta)
        }
        e <- (A %*% theta - b)
        d <- (t(A) %*% e) / length(b)
        oldtheta <- theta
        theta <- theta - alpha * d
    }

    return(theta)
}
```

R Function 6.10
Least squares via gradient descent


```
[2,]    1  3.0  1.4  0.2
[3,]    1  3.2  1.3  0.2
[4,]    1  3.1  1.5  0.2
[5,]    1  3.6  1.4  0.2
[6,]    1  3.9  1.7  0.4
```

We would also like to test the model. For that, we will use the lm function, and include "- 1" in the specification. This is because the matrix A includes a column of 1 to support finding an intercept, so the lm function will not add one of its own to the linear model.

```
> lm(b ~ A - 1, data = iris)
Call:
lm(formula = b ~ A - 1, data = iris)

Coefficients:
     A1        A2        A3        A4
 1.8560    0.6508    0.7091   -0.5565
```

These are the coefficients, for the intercept, sepal width, petal length, and petal width, respectively. Using gdls with a maximum of 100 iterations, we can see the results are not particularly close.

```
> gdls(A, b, alpha = 0.05, m = 100)
            [,1]
```

```
[1,]   0.43272276
[2,]   1.08614868
[3,]   0.57378984
[4,]  -0.07166668
```

In fact, the results might be considered discouraging. But increasing the maximum number of steps yields a result much closer to the results from lm:

```
> gdls(A, b, alpha = 0.05, m = 1000)
            [,1]
[1,]   1.0185004
[2,]   0.8718602
[3,]   0.7728138
[4,]  -0.6284752
> gdls(A, b, alpha = 0.05, m = 10000)
            [,1]
[1,]   1.8439696
[2,]   0.6538332
[3,]   0.7107731
[4,]  -0.5593274
```

However, like concerns about the step size in gd, the gdls function is sensitive to changes in its learning rate.

```
> gdls(A, b, alpha = 0.1, m = 1000)
              [,1]
[1,]  -1.243656e+253
[2,]  -3.757973e+253
[3,]  -5.297415e+253
[4,]  -1.755591e+253
```

That result is clearly incorrect and with a larger maximum number of iterations, would overflow. As a rule, gradient descent is substantially slower than other methods of finding the least squares solution. However, as a template, it stands out. Other forms from the generalized linear model framework can be implemented by changing the cost function and its associated derivative.

6.4.2 The Traveling Salesperson

The traveling salesperson[3] problem is a well-known problem from both computer science and numerical analysis. In the problem, a salesperson is to visit a series of distinct cities, $C_1, C_2, \ldots, C_n$ and return the first city. To save time and money, the salesman would like to find the most efficient path that hits each city before getting home. Extraneous time spent in transit is lost.

It is possible that our salesperson can check every possible option, but that is a task that becomes substantially more difficult as cities are added.

---

[3]This problem is historically referred to as the "traveling salesman problem," but it is 2017 now.

The total number of potential paths is $n!$ for n cities to visit. For $n = 5$, there are $5! = 120$ potential paths, a number low enough that it could be checked with a simple spreadsheet or even by hand over the course of an afternoon. However, for $n = 10$, there are $10! = 3628800$ options and at $n = 32$, there are $32! \approx 2.63 \times 10^{35}$ potential options. The number of potential paths moves from computationally feasible to impractical as quickly as it goes from pen and paper to a computer.

The traveling salesperson problem is very well suited for use with simulated annealing. We will start with a list of cities and their locations. We will use runif, one of R's random number generators, to create a list of 30 integers over $[0, 99]$. These are divided into 15 groups of two that provide x and y coordinates for each of the cities on an imaginary map. The coordinates are given as integers for aesthetic reasons and provides no computational purpose.

```
> head(cities <- matrix(floor(runif(30, 0, 100)), ncol = 2))
     [,1] [,2]
[1,]   68   92
[2,]   37   71
[3,]    4   73
[4,]   76   57
[5,]    0   22
[6,]   98   57
```

As we can see from the first few lines of the cities list, the coordinates are simply specified.

Simulated annealing starts with an initial vector describing a single potential path through these cities. However, just as in the continuous case, there is no need to necessarily pick a path that is very good. So we will set the initial path to visit each city in the list in order and return to the first. In the place of a function evaluation, we calculate the distance, using vecnorm, between each step. Finally, in order to create a random change, we pick a single city at random from the list, and switch it with the city after it in the order. If the route is more efficient, we keep it. If not, we may keep it anyway, subject to the temperature control.

This change in simulated annealing is written as a custom traveling salesperson implementation, given in Function 6.11. The tspsa function accepts as input the cities list, as an n column vector. Using an n column vector allows for more interesting problems, such as an n-dimensional traveling sales person problem, to be solved. The function also accepts an initial temperature and cooling rate. The return value is a list with the best found path and its associated distance.

At the start, we can set the temperature to 1. In the initial step, the loop will not evaluate and no alternative paths are taken. This has the side effect of returning the distance of the path for visiting the cities in order.

```
> tspsa(cities, temp = 1)
```

```
tspsa <- function(x, temp = 1e2, rate = 1e-4) {
    step = 1 - rate
    n <- nrow(x)

    xbest <- xcurr <- xnext <- c(1:n)
    ynext <- 0
    for(i in 2:n) {
        a <- xnext[i - 1]
        b <- xnext[i]
        ynext <- ynext + vecnorm(x[a,] - x[b,])
    }
    a <- xnext[1]
    b <- xnext[n]
    ynext <- ynext + vecnorm(x[a,] - x[b,])
    ybest <- ycurr <- ynext

    while(temp > 1) {
        temp <- temp * step
        i <- ceiling(runif(1, 1, n))
        xnext <- xcurr
        temporary <- xnext[i]
        xnext[i] <- xnext[i - 1]
        xnext[i - 1] <- temporary
        ynext <- 0
        for(i in 2:n) {
          a <- xnext[i - 1]
          b <- xnext[i]
          ynext <- ynext + vecnorm(x[a,] - x[b,])
        }
        a <- xnext[1]
        b <- xnext[n]
        ynext <- ynext + vecnorm(x[a,] - x[b,])
        accept <- exp(-(ynext - ycurr) / temp)
        if(ynext < ycurr || runif(1) < accept) {
            xcurr <- xnext
            ycurr <- ynext
        }
        if(ynext < ybest) {
            xbest <- xcurr
            ybest <- ycurr
        }
    }
    return(list(order = xbest, distance = ybest))
}
```

R Function 6.11
The traveling salesperson problem via simulated annealing

```
$order
 [1]  1  2  3  4  5  6  7  8  9 10 11 12 13 14 15
```

```
$distance
[1] 934.1485
```

From here, we have a bad scenario, but it is probably not the worst-case scenario. We are equally unlikely to have randomly picked the worst path as the best. So we will use **tspsa**, along with a more reasonable initial temperature to find a better path through the cities.

```
> tspsa(cities, temp = 100)
$order
 [1] 12  2  8  6 13  4 10 11  9 14  5 15  7  3  1
```

```
$distance
[1] 369.4153
```

That option is substantially better and around half of the initial vector. But the important thing to remember about simulated annealing is that the result found will not necessarily be the best. It will only be somewhere near the best option. Repeating the analysis three more times will yield different routes with different results, though they have similar total distances. This is true even though each attempt to optimize the path begins with the same initial path.

```
> tspsa(cities, temp = 100)
$order
 [1]  5 14  3  2 12  1  8  6 13  4 10 11  9  7 15
```

```
$distance
[1] 384.9854
> tspsa(cities, temp = 100)
$order
 [1]  7 15  5 14  2 12  1  8  6 13  4 10 11  9  3
```

```
$distance
[1] 391.547
> tspsa(cities, temp = 100)
$order
 [1]  3  2 12  1  8  6 13  4 10 11  9 14  5 15  7
```

```
$distance
[1] 316.5385
```

Increasing the initial temperature will have some effect, generally, helping **tspsa** produce slightly better results. However, the tradeoff is spectacular given the temperature increase over 100 and the results will not necessarily be better. The initial temperature selection has the effect of setting the initial jump sizes used within the simulated annealing process.

```
> tspsa(cities, temp = 10000)
$order
 [1]  6  4 10 11  9 14  5 15  7  3  2  1 12 13  8

$distance
[1] 342.2154
> tspsa(cities, temp = 10000)
$order
 [1] 14  5 15  7  3  2 12  1  8  6 13  4 10 11  9

$distance
[1] 316.5385
> tspsa(cities, temp = 10000)
$order
 [1]  6 12 11  9 14  5 15  7  3  2 13  4  1 10  8

$distance
[1] 431.7529
```

There is also an option, *rate* that defines the rate at which the temperature decreases. The default is 1/10000.

```
> tspsa(cities, temp = 100, rate = 1e-2)
$order
 [1]  7  2  1  8  6 12  3 11 10  4 13  9 14  5 15

$distance
[1] 483.3175
> tspsa(cities, temp = 100, rate = 1e-4)
$order
 [1] 10  8  6 13 11  9 14  5 15  7  3  2 12  1  4

$distance
[1] 326.0436
> tspsa(cities, temp = 100, rate = 1e-6)
$order
 [1]  7 15  5 14  9 11 10  4 13  6  8  1 12  2  3

$distance
[1] 316.5385
```

Higher temperatures and lower rates increase the number of times through the loop and increase the time of calculation. However, the results become more reliable, meaning we can trust the smallest path has been found, when more calculations are made.

As we can see, we have used the discrete version of simulated annealing to solve an important problem. Variations on this problem turn up in logistics and planning contexts, so having a solid solution is important. For even moderately large problems, such as 15 cities, using an exhaustive search is likely the best solution. While time consuming, the results are exact. However, as the number of cities increases, finding probabilistic solvers will make finding good enough solutions possible.

Comments

Optimization is a critical problem in both mathematics and computer science with applications in physics, statistics, economics, and many other fields. Accordingly, optimization is a frequently studied problem. This chapter has only touched on a handful of methods. Entire books can and have been written only on optimization and especially on application spaces.

Within R, there is a built-in function, uniroot, which excels at finding the the root of one-dimensional functions. The uniroot function itself accepts a function, a list containing a lower bound and an upper bound of the domain to search for a root, as well as a maximum tolerance for error. The uniroot function is implemented internally using Brent's method, which combines the binary search of the bisection method with the rapid convergence of Newton–Raphson. We can compare the results of uniroot to our bisection function using the cubic formula, $f(x) = (x - 2)^3 - 1$:

```
> f <- function(x) { (x - 2)^3 - 1 }
> bisection(f, 0, 10)
[1] 3.000183
> uniroot(f, c(0, 10))
$root
[1] 3

$f.root
[1] 5.401947e-07

$iter
[1] 12

$init.it
[1] NA
```

```
$estim.prec
[1] 6.103516e-05
```

The uniroot function takes a different argument structure, has a lower default tolerance, and provides diagnostics information in the return. However, the method and result are comparable to bisection.

In addition to uniroot, R also provides the function polyroot for finding multiple roots of a polynomial function. Polynomial functions are specified, like in Section 1.3.2, as an array of coefficients in increasing order. Therefore, the function $f(x) = 3x^2 + 2x + 1$ would be specified as an array c(3, 2, 1), for which polyroot will find the roots. For the trivial example of $f(x) = x^2 - 1$, which has roots at 1 and -1, we can see the results:

```
> polyroot(c(1, 0, -1))
[1]  1+0i -1+0i
```

As we can see, the results, instead of a vector of real numeric values, is a vector of complex numbers. The polyroot function is capable of resolving complex roots, as the trivial example of $f(x) = x^2 + 1$ shows:

```
> (roots <- polyroot(c(1, 0, 1)))
[1] 0+1i 0-1i
```

which correctly returns the result, $0 \pm i$. The real and imaginary components of the complex data type can be extracted using the Re and Im functions, respectively.

```
> Re(roots)
[1] 0 0
> Im(roots)
[1]  1 -1
```

The polyroot function can breakdown on high degree polynomials. The real power of polyroot comes in the ability to find the roots of polynomials approximating some other function. As we can recall from Chapters 4 and 5, polynomials are used to approximate unknown and known nonpolynomial functions for the purposes of interpolation and integration.

Fittingly, there are also a number of R packages and functions dedicated to optimization. First, the optim function in the stats package provides a generalized interface to optimization. Both the optimx and ROI packages provide infrastructure for many different kinds of optimization. Simulated annealing is provided by several packages, including the GenSA package. Even the traveling salesperson problem has a dedicated package, TSP.

Before solving an optimization problem in R, we should carefully consider the optimization task view inside CRAN. This task list provides far more packages than those listed here and will likely have an optimization method suitable for our needs.

Exercises

1. Find a real root of the function $f(x) = x^3 - 2x + 2$ using the Newton–Raphson method and $x_0 = 0$. Also try with $x_0 = 1/2$.

2. Reimplement the golden section search as a recursive algorithm.

3. Reimplement the Newton–Raphson method for root finding as a recursive algorithm.

4. What advantage does the bisection method provide over the Newton–Raphson method for root finding?

5. Which of the bisection, Newton–Raphson, or the secant method can be used to find the real root of a nonpolynomial function?

6. Given $f(x) = \sin x/x$, list all real roots between $x = 0.5$ and $x = 1$.

7. Under what conditions is the secant method preferable to the Newton–Raphson method for root finding?

8. A standard example of a test case is known as the Wilkinson polynomial, $W(x)$, such that $W(x) = (x - 1)(x - 2) \cdots (x - 20)$. At first inspection, there is a real root at each integer between 1 and 20, inclusive. However, the Wilkinson polynomial is known to cause difficulty for numerical algorithms. Implement Wilkinson's polynomial and show an approach for finding all of the roots, if none were known.

9. Why does the golden section search fail to work with multidimensional functions?

10. How can gradient descent be extended to find a global minimum or maximum?

11. What advantages does gradient descent provide over hill climbing? What advantages does hill climbing provide over gradient descent?

12. Is either hill climbing or gradient descent guaranteed to locate a global extreme? Explain.

13. Using the information included in this chapter, implement a function that will minimize or maximize the solutions of systems of equations.

14. We can implement logistic regression, sometimes called "logit" using the method shown in Section 6.4.1. The cost function for logistic regression is,

$$J(\theta_1, \theta_2, \ldots) = \frac{1}{m} \sum_{i=1}^{m} [-y_i \log{(h_\theta(x_i))} - (1 - y_i) \log{(1 - h_\theta(x_i))}]$$

$$(6.13)$$

where $h_\theta(x)$ is the logistic function,

$$h_\theta(x) = \frac{1}{1 + e^{-\theta^\intercal x}}. \tag{6.14}$$

As with least squares, the cost function is only used for finding the mean squared error, and the function is minimized using the derivative,

$$J'(\theta) = \frac{1}{m} X^\intercal (h_\theta(x) - y). \tag{6.15}$$

Implement logistic regression.

15. Find a list of the Major League Baseball stadia and their locations. Use simulated annealing to find an optimal path to visit each one.

7

Differential Equations

7.1 Initial Value Problems

Initial value problems are the first class of differential equation problems we are interested in solving. Ordinary differential equations are complicated by the fact they do not provide sufficient information to solve them, normally. A differential equation, $f'(x, \ldots)$, is the result of differentiating some other function, $f(x, \ldots)$. The process of solving the differential equation, and finding a value of $f(x, \ldots)$ for some value of $x, \ldots$, is not possible because the integral of $f'(x, \ldots)$ merely describes the general shape. The vertical shift, up or down, is not known. This vertical shift is the constant of integration.

During differentiation, the value of whatever vertical shift is present is lost as a result of the elimination of the constant term, which has a derivative of 0. We normally acknowledge this when integrating a function by adding a $+C$ constant, the constant of integration, to an indefinite integral. This is sometimes a nonissue since, if finding the value of a definite integral, the constant terms cancel and the constant of integration is unnecessary.

For ordinary differential equations, there is no convenient cancellation, leading to the initial value problem. The initial value problem provides a value of $f(x_0, \ldots)$, where x_0 is normally 0, but is not required to be. This initial value provides sufficient information to complete the solution and find the actual value of $f(x, \ldots)$ for some value of x. This section provides several methods for solving initial value problems numerically.

7.1.1 Euler Method

The elementary method of solving initial value problems is the Euler method. With the Euler method, we follow the same intuitive approach we use to initially approach other problems, numerically. Assuming we have a differential equation, $f'(x, \ldots)$, we are interested in finding the value of $f(x, \ldots)$, and we know that there is a relationship between these two equations. The value of $f'(x, \ldots)$ is the slope of $f(x, \ldots)$ at $x, \ldots$. We can see this in action in the vector field in Figure 7.1.

If we are willing to assume that the function is linear over the short run, we can estimate the value of $f(x)$ by adding the change in y to $f(x)$. We can

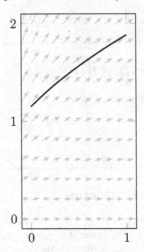

Figure 7.1
Vector field of $f'(x, y) = \frac{y}{2x+1}$

find the change in y by multiplying the slope at x times the change in x. This is essentially the same process as the finite differences method for estimating the derivative, except applied in reverse. In brief,

$$f(x + h) \approx f(x) + hf'(x, y), \tag{7.1}$$

where h is the step size.

Provided the change in x, the step size, is sufficiently small, $f(x)$ will be sufficiently linear that the estimate is suitable. For an infinitesimally small step size, the result is precise, but as we know, there are limits to how much storage space is available within the machine and finding the value of $f(x)$ at some $x > x_0$ would require an infinite number of steps. With the Euler method, we specify a step size, the starting value of x_0, and the starting value of y_0, as well as the far end point, $x^\star$, for which we wish to find $f(x^\star)$. This logic is implemented in Function 7.1, the `euler` method, which does not specify the second endpoint, but does allow the user to specify the number of steps taken. Therefore, $x^\star = x_0 + hn$, where n is the total number of steps taken.

Like most of the methods we have seen, the Euler method is more accurate as the resolution is increased, by increasing the value of n. And like most other methods, the computation becomes more intense as the resolution is increased. For n steps, the estimate requires n evaluations of $f'(x)$. In addition, the smaller step size risks storage problems as a small $x_i + h$ may not be distinguishable from x_i. The larger concern about the Euler method is the algorithmic error. The Euler method functions based on the assumption that $f(x)$ is approximately linear over the domain $[x, x + h]$, an assumption

```
euler <- function(f, x0, y0, h, n) {
    x <- x0
    y <- y0

    for(i in 1:n) {
        y0 <- y0 + h * f(x0, y0)
        x0 <- x0 + h
        x <- c(x, x0)
        y <- c(y, y0)
    }

    return(data.frame(x = x, y = y))
}
```

R Function 7.1
Euler method for ordinary differential equations

that is not necessarily true. However, as h decreases, the approximately linear assumption is generally more valid.

For example, we will consider $f'(x, y)$ such that,

$$f'(x,y) = \frac{y}{2x+1}. \tag{7.2}$$

Because this is an example, we know that given $f(0) = 1$,

$$f(x) = \sqrt{2x+1}. \tag{7.3}$$

We also know that $f(1) = \sqrt{3} \approx 1.732051$. Given this example, we can examine the performance of the Euler method. First, the Euler method with $h = 0.5$ and $n = 2$:

```
> f <- function(x, y) { y / (2 * x + 1) }
> euler(f, 0, 1, 1 / 2, 2)
    x    y
1 0.0 1.000
2 0.5 1.500
3 1.0 1.875
```

The function returns a data frame containing the calculated x and y values, including the initial value. The example again with $h = 0.2$ and $n = 5$:

```
> euler(f, 0, 1, 1 / 5, 5)
    x        y
1 0.0 1.000000
2 0.2 1.200000
3 0.4 1.371429
4 0.6 1.523810
```

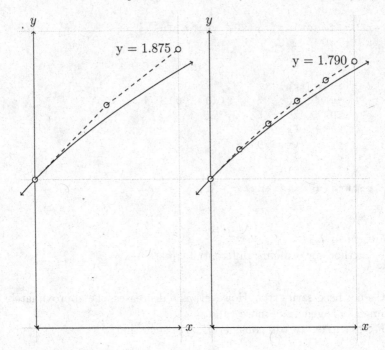

Figure 7.2
Euler method with $h = 0.5$ and 0.2

```
5 0.8 1.662338
6 1.0 1.790210
```

With the smaller h step size and more steps, the method is substantially closer. We can see this in Figure 7.2, where with $h = 0.2$, the projected value of y remains closer to the true value over the lifetime of the integration, versus a larger step size.

Performance continues to improve with smaller step sizes. We can use the `tail` function to limit the display to just the last line, the calculated value of $f(1)$:

```
> tail(euler(f, 0, 1, 1 / 10, 10), 1)
    x       y
11 1 1.76109
> tail(euler(f, 0, 1, 1 / 100, 100), 1)
     x       y
101 1 1.73494
> tail(euler(f, 0, 1, 1 / 1000, 1000), 1)
      x       y
1001 1 1.73234
```

The calculated value of $f(1)$ steadily converges to $\sqrt{3}$ as h decreases. With initial value problems, we can define our algorithmic error in two ways. The first is local truncation error. This is the error from one step (at x_i) to the next (at $x_i + h$). Local truncation error exists with an embedded assumption that the estimate at x_i is correct. At x_0, where y_0 is the initial value we set, this is reasonable. However, each step adds local truncation error to the last. The accumulated local truncation error is called global truncation error.

Our algorithms for solving initial value problems should minimize both types of error. Given that all of our problems have simple solutions that are somewhat error prone, it should be no surprise that the straightforward approach to initial value problems is subject to some degree of error. Like the Newton–Cotes methods of integration, the Euler method approximates the original function, $f(x)$, using the local properties of $f'(x)$, and the methods present similar errors. The local truncation error, over $[x_i, x_i + h]$, is

$$f(x_i + h) - f(x_i) = \frac{h^2 f^{(2)}(\xi)}{2}, \tag{7.4}$$

where $\xi \in [x_i, x_i + h]$. Like the Newton–Cotes methods, the local error is dependent upon both the second derivative of $f(x)$ and the step size. For a sufficiently uninteresting differential equation, where $f(x)$ is linear, the Euler method is precise, as $f^{(2)}(\xi) = 0$. We can also see this geometrically as each step is a straight shot along the straight line described by $f'(x)$.

For more interesting differential equations, the local truncation error will not reduce to 0, and the error with each step grows. Over the domain from $[x_0, x_n]$, there are n steps. For each of these steps, there is an error bounded by h^2, per the local truncation error. Accordingly, the global truncation error is proportional to $h^2/h = h$.[1]

Some differential equations are numerically unstable with standard evaluation tools and reasonable step sizes. We call these equations stiff, but there is no standard definition nor test for stiffness, which leads to stiff equations occurring spontaneously in practice. In general, an equation is considered stiff if small changes in the step size used in a method like Euler lead to dramatic changes in the results. In some sense, stiffness is a matter of opinion, though usually an easily agreeable one.

An entirely contrived example of a stiff ordinary differential equation where $f'(x, y) = -10y$ is shown in Figure 7.3. For the Euler method with a step size of $h = 1/5$, the resultant points oscillate between -1 and 1. However, $f(x) = \exp{-10x}$, a function that rapidly decays to zero with no oscillation. The function is shown in the figure, along with the results of the Euler method with a step size of $h = 1/5$. As we can see, the projected values of the function bounce between 1 and -1, subject to some roundoff error.

There are two ways to address stiff differential equations. The first way

---

[1]There is a formal specification of the error bounds, but the details are not critical. Understanding the proportion, as a function of h, is more useful and provides a suitable shorthand.

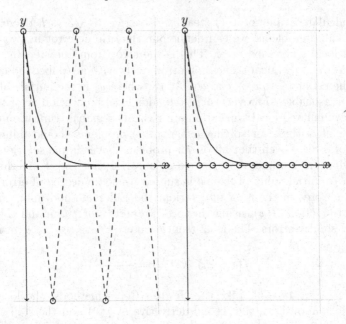

Figure 7.3
Integration of $f'(x, y) = -10y$

is to reduce to the step size used in the method. Reducing the step size sufficiently should overcome the stiffness leading to an appropriately modeled
differential equation. An additional advantage is that the overall error is reduced. However, the penalty for processing stiff ordinary differential equations
is the increased number of function evaluations necessary to complete the integration. Further, it is possible the given function will require such a fine step
size that calculation is unreasonably small. At a smaller step size, the process
may require too many calculations to be feasible.

The second way is to improve the calculation method used. Similar to the
process used in numerical integration, that uses improved estimates instead
of just more evaluation points, we can use more detailed calculations to estimate the value of the differential equation. These methods are generally called
Runge–Kutta methods and the Euler method shown here is the most elementary form of them. More complex and more interesting estimates are available
for stiff or more difficult differential equations.

We can measure the effectiveness of a differential equations method on stiff
differential equations. This is measured using a concept known as A-stability.
A differential equation solver is considered A-stable based on the results of its
application to the differential equation,

$$f'(x, y) = kx, \tag{7.5}$$

for some value of k. The analytical solution to this differential equation is $f(x) = \exp kx$ and as x increases, then the value of $f(x)$ goes to 0 if $k < 0$. This is true regardless of step size used. So if for a fixed step size, a numerical method also approaches 0, the method can be called A-stable.

As we saw in the example $f'(x, y) = -10y$, the Euler method is not A-stable because it is not stable at a step size of $h = 1/5$. The method may seem stable at smaller step sizes, but we are better off finding an algorithm with better stability. We will see some examples of other algorithms going forward that can manage this test better. After Runge–Kutta methods, we will look further toward even smarter methods, such as Adams–Bashforth, that use weighted steps to increase the precision of the intermediate results. All of these are usable on stiff ordinary differential equations, at the cost, again, of increased complexity and an increased number of calculations.

7.1.2 Runge–Kutta Methods, Generally

The Euler method is the elementary form of what are broadly called Runge–Kutta methods. Runge–Kutta methods, like Newton–Cotes methods, use an increasing number of calculation points to estimate the position of a function from its first derivative.

However, for a curve that is changing direction over the step's domain $[x_i, x_{i+1}]$, the derivative of the function $f(x)$ at x_i may not accurately capture the derivative over the entire step. Since our entire basis for understanding $f(x)$ is its first derivative, we can assume $f(x)$ is differentiable and therefore continuous, so we can estimate the value of $f(x)$ anywhere along the step, not only at the endpoints of the step. If the function is relatively smooth over the entire step, we can infer that the derivative at the midpoint is closer to the "average" derivative over the step. This inference is instinctive, but for smooth differential equations, the inference holds well.

We can see the difference in action in Figure 7.4. The figure on the left shows the Euler method where $f(x) = -2(x-1)^2 + 1$. Using the Euler method, shown at left, overshoots the correct value at x_{i+h} by quite a bit. Using this half step method, shown in on the right of Figure 7.4, uses the slope at $x_i + h/2$, shown in line b, to jump from the value at x_i, shown in line a, to estimate $x_i + h$, which is much, much closer.

Using this inference, we can take a half-step to the midpoint of the domain $[x_i, x_{i+1}]$ and using the estimate of $f(x)$ at the midpoint, take a full step to the end of the domain. This method, called `midptivp`, is provided in Function 7.2.

Like for the Euler method example, we will again consider $f'(x, y)$ such that,

$$f'(x, y) = \frac{y}{2x + 1}, \tag{7.6}$$

where $f(1) = \sqrt{3} \approx 1.732051$. Using $h = 0.5$ and $n = 2$:

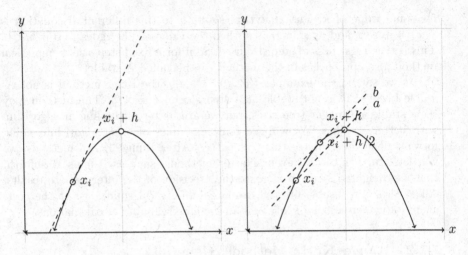

Figure 7.4
Euler and midpoint initial value problem example

```
> f <- function(x, y) { y / (2 * x + 1) }
> midptivp(f, 0, 1, 1 / 2, 2)
    x        y
1 0.0 1.000000
2 0.5 1.416667
3 1.0 1.735417
```

The function returns a data frame containing the calculated x and y values, including the initial value. The example again with $h = 0.2$ and $n = 5$:

```
> midptivp(f, 0, 1, 1 / 5, 5)
    x        y
1 0.0 1.000000
2 0.2 1.183333
3 0.4 1.341815
4 0.6 1.483452
5 0.8 1.612692
6 1.0 1.732314
```

Like the Euler method, the midpoint method performance continues to improve with smaller step sizes.

```
> tail(midptivp(f, 0, 1, 1 / 10, 10), 1)
    x        y
11 1 1.732085
> tail(midptivp(f, 0, 1, 1 / 100, 100), 1)
```

```
midptivp <- function(f, x0, y0, h, n) {
    x <- x0
    y <- y0

    for(i in 1:n) {
        s1 <- h * f(x0, y0)
        s2 <- h * f(x0 + h / 2, y0 + s1 / 2)
        y0 <- y0 + s2

        x0 <- x0 + h
        x <- c(x, x0)
        y <- c(y, y0)
    }

    return(data.frame(x = x, y = y))
}
```

R Function 7.2

Midpoint method for ordinary differential equations

```
      x        y
101   1   1.732051
```

As opposed to the Euler method, the midpoint achieved accuracy within the number of printed digits, with a step size of only 1/100. Further decreases in h will gain additional accuracy in the final result and selecting an appropriate step size is subject to the application's requirements.

Like many of our routines, we need a set of assumptions to align before making an application suitable. The half step assumption here requires a certain level of smoothness to the underlying function. If $f(x)$ oscillates widely over $[x_i, x_i + h]$, then this method is far less effective. More often, those functions are sufficiently smooth so that this sort of method can work. As a family, these Runge–Kutta methods use more and more intermediate steps to estimate the value of $f(x_i + h)$. The most important of these is the fourth-order Runge–Kutta method, which uses four intermediate steps, shown in Function 7.3, the **rungekutta4** function.

Using equation 7.6 as the differential equation and $h = 0.2$ and $n = 5$:

```
> f <- function(x, y) { y / (2 * x + 1) }
> rungekutta4(f, 0, 1, 1 / 2, 2)
     x        y
1  0.0   1.000000
2  0.5   1.414931
3  1.0   1.732995
```

The function returns a data frame containing the calculated x and y values, including the initial value. The example again with $h = 0.2$ and $n = 5$:

```
rungekutta4 <- function(f, x0, y0, h, n) {
    x <- x0
    y <- y0

    for(i in 1:n) {
        s1 <- h * f(x0, y0)
        s2 <- h * f(x0 + h / 2, y0 + s1 / 2)
        s3 <- h * f(x0 + h / 2, y0 + s2 / 2)
        s4 <- h * f(x0 + h, y0 + s3)
        y0 <- y0 + s1 / 6 + s2 / 3 + s3 / 3 + s4 / 6

        x0 <- x0 + h
        x <- c(x, x0)
        y <- c(y, y0)
    }

    return(data.frame(x = x, y = y))
}
```

R Function 7.3
Fourth-order Runge–Kutta method for ordinary differential equations

```
> rungekutta4(f, 0, 1, 1 / 5, 5)
    x        y
1 0.0 1.000000
2 0.2 1.183234
3 0.4 1.341666
4 0.6 1.483270
5 0.8 1.612485
6 1.0 1.732087
```

Like the Euler method, the midpoint method performance continues to improve with smaller step sizes.

```
> tail(rungekutta4(f, 0, 1, 1 / 10, 10), 1)
     x        y
11 1 1.732053
> tail(rungekutta4(f, 0, 1, 1 / 100, 100), 1)
      x        y
101 1 1.732051
```

The fourth-order Runge–Kutta method is widely considered the standard approach to numerical solutions of initial value problems. As a rule, when "Runge–Kutta" is named, but the parameters are unspecified, the speaker is almost certainly referring to the fourth-order method. The method is included in the pracma library, where it is called rk4 and accepts slightly different arguments. Instead of specifying the initial starting point, the step size, and

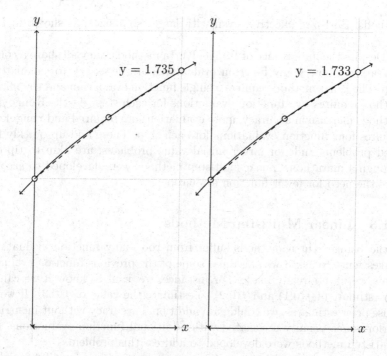

Figure 7.5
Midpoint and fourth order Runge–Kutta methods with $h = 0.5$

number of steps, like **rungekutta4**, the **rk4** function accepts the endpoints along with the number of steps. But the result is the same.

```
> library(pracma)
> rk4(f, 0, 1, 1, 5)
$x
[1] 0.0 0.2 0.4 0.6 0.8 1.0

$y
[1] 1.000000 1.183234 1.341666 1.483270 1.612485 1.732087
```

The rk4 function returns its values as two lists rather than a data frame.

As we also noted with the Euler method, neither the midpoint method nor the fourth-order Runge–Kutta method's errors are critical to understanding the procedure. What is important is understanding how the error changes with respect to the step size. For the midpoint method, the local truncation error is proportional to h^3 and the global truncation error is proportional to h^2. For the fourth-order Runge–Kutta method, the local truncation error is proportional to h^5 and the global truncation error is proportional to h^4. For smaller step sizes, both error types decrease more rapidly for higher-order methods. Both are effective, even with larger step sizes, as shown in Figure 7.5.

Despite the importance of Runge–Kutta methods, they all share a common flaw around how many function evaluations are necessary to calculate each step. The Euler method requires a single function evaluation and the midpoint method requires two function evaluations for each step. As the Runge–Kutta methods increase in accuracy, more evaluations are required and `rungekutta4` requires four function evaluations for each step. These add up quickly in applied problems and, for larger initial value problems, are akin to tip-toeing through a marathon. Linear multistep methods were developed to accommodate the need for fewer function evaluations.

7.1.3 Linear Multistep Methods

If the Runge–Kutta methods suffer from too many function evaluations, it makes sense to ask if we can reuse some of the previous function evaluations, those we have already made. For instance, we want to know if we can reuse the estimate of $f(0.1)$ and $f(0.2)$ to estimate the value of $f(0.3)$. If we could reuse prior estimates, we could gain additional accuracy without incurring the performance penalty associated with additional function evaluations. Linear multistep methods were developed to address this problem.

In one sense, the elementary linear multistep method for differential equations includes only one point, x_i, in the calculation of x_{i+1}. This is exactly how the Euler method functions and the Euler method is the elementary linear multistep method. The next method uses x_{i-1} and x_i to calculate x_{i+1}. The Adams–Bashforth methods use a weighted addition, including negative weights, of steps and points to arrive at the next step. Like other numerical methods, the weights arise from the polynomial interpolation of the points available.

The second-order Adams–Bashforth method is based on the formula,

$$f(x_{i+2}) = f(x_{i+1}) + \frac{h}{2}(3f'(x_{i+1}) - f'(x_i)). \tag{7.7}$$

This approach interpolates between the preceding points to estimate the third point in the group. This third point becomes the midpoint of the next iteration as the entire process advances. Because the previous value is stored and reused, an additional function evaluation is unnecessary. If the Runge–Kutta methods could be compared to tip-toeing through the vector field, then the Adams–Bashforth methods could be equally compared to running through the vector field. This does not, however, mean the Adams–Bashforth methods are superior.

This approach, for two preceding points, is shown in `adamsbashforth`, given in Function 7.4. The second-order Adams–Bashforth method requires two initial values to begin, x_0 and x_1. However, an initial value problem only provides a value of $f(x_0)$.

```
adamsbashforth <- function(f, x0, y0, h, n) {

    ## Quick Euler the value of x1, y1
    y1 <- y0 + h * f(x0, y0)
    x1 <- x0 + h

    x <- c(x0, x1)
    y <- c(y0, y1)
    n <- n - 1

    for(i in 1:n) {
        yn <- y1 + 1.5 * h * f(x1, y1) - .5 * h * f(x0, y0)
        xn <- x1 + h

        y0 <- y1
        x0 <- x1
        y1 <- yn
        x1 <- xn

        y <- c(y, y1)
        x <- c(x, x1)
    }

    return(data.frame(x = x, y = y))
}
```

R Function 7.4
Adams–Bashforth method for ordinary differential equations

It is possible the value of $f(x_1)$ could be provided at the problem instantiation. However, the common approach is to use another method to take the initial step. In this implementation, an embedded Euler step provides the value of $f(x_1)$, due to its relative simplicity. For a higher-order Adams–Bashforth implementation, it would be better to use a higher-order Runge–Kutta method, likely the fourth-order method, to take the initial step. However, higher-order embedded first steps require more complex implementations.

The implementation of adamsbashforth is itself simplistic, like the implementations of Euler and the Runge–Kutta methods. The variables for storing prior values are recycled in the interest of economy of implementation. Adams–Bashforth methods of different orders suffer from a common problem of initial data.

We can see the second-order Adams–Bashforth method in action and compare it to other initial value problem solvers we have seen. We will again consider $f'(x, y)$ such that,

$$f'(x, y) = \frac{y}{2x + 1}, \tag{7.8}$$

where $f(1) = \sqrt{3} \approx 1.732051$. Using $h = 0.5$ and $n = 2$:

```
> f <- function(x, y) { y / (2 * x + 1) }
> adamsbashforth(f, 0, 1, 1 / 2, 2)
    x    y
1 0.0 1.0000
2 0.5 1.5000
3 1.0 1.8125
```

The function returns a data frame containing the calculated x and y values, including the initial value. The example again with $h = 0.2$ and $n = 5$:

```
> adamsbashforth(f, 0, 1, 1 / 5, 5)
    x        y
1 0.0 1.000000
2 0.2 1.200000
3 0.4 1.357143
4 0.6 1.497619
5 0.8 1.626443
6 1.0 1.746036
```

With the smaller h step size and more steps, the method is substantially closer. We can see this in Figure 7.6, where with $h = 0.2$, the projected value of y remains closer to the true value over the lifetime of the integration, versus a larger step size.

Performance continues to improve with smaller step sizes. We can use the `tail` function to limit the display to just the last line, the calculated value of $f(1)$:

```
> tail(adamsbashforth(f, 0, 1, 1 / 10, 10), 1)
   x        y
11 1 1.735717
> tail(adamsbashforth(f, 0, 1, 1 / 100, 100), 1)
    x        y
101 1 1.732089
> tail(adamsbashforth(f, 0, 1, 1 / 1000, 1000), 1)
     x        y
1001 1 1.732051
```

The function estimate will converge to the correct answer as h approaches zero. The second-order Adams–Bashforth method, as a second-order method, has a global truncation error that is proportional to h^2. Accordingly, it converges faster than the Euler method and requires half as many function evaluations as the midpoint method for initial value problems. An example of the convergence is shown in Function 7.6. Like higher-order Runge–Kutta methods, Adams–Bashforth methods can be used to address stiff ordinary differential equations.

There are additional Adams–Bashforth methods for higher-order solutions.

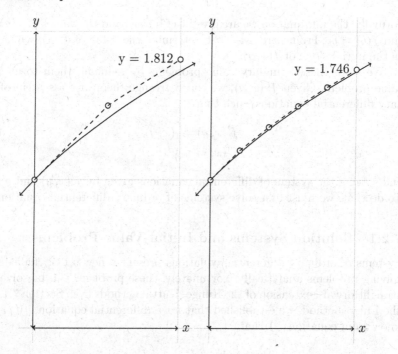

Figure 7.6
Adams–Bashforth with $h = 0.5$ and 0.2

For instance, the third-order Adams–Bashforth method relies on the function,

$$f(x_{i+3}) = f(x_{i+2}) + \frac{h}{12}(23f'(x_{i+2}) - 16f'(x_{i+1}) + 5f'(x_i)). \qquad (7.9)$$

An implementation of equation 7.9 would yield a third-order initial value problem solver, with global truncation error proportion to h^3, and still only requiring n function evaluations. Such an implementation would also need a reasonable way to generate the first two estimates, x_1 and x_2, but the Runge–Kutta methods can be used there in the same way as in the second-order method.

7.2 Systems of Ordinary Differential Equations

Boundary value problems are another type of problem introduced by ordinary differential equations. In a boundary value problem, we are presented with information about the endpoints of our function and its *second* derivative.

Formally, the information we are given is $f''(x, y)$, and the values of $f(a) = y_a$ and $f(b) = y_b$. From here, we must determine the value of $f'(a)$, which gives us the initial slope of $f(x, y)$.

We solve these boundary value problems by reducing them to an initial value problem. Given $f''(x, y)$, we can rearrange them into a system of ordinary differential equations, such that,

$$f'(x, y) = g(x, y) \qquad\qquad (7.10)$$
$$g'(x, y) = f''(x, y), \qquad\qquad (7.11)$$

and solving the system of differential equations given by $f'(x, y)$ and $g'(x, y)$. To do that, we must first solve systems of ordinary differential equations.

7.2.1   Solution Systems and Initial Value Problems

Systems of ordinary differential equations present a new set of challenges for solving problems analytically. Fortunately, these problems can be solved in a straightforward extension of the Runge–Kutta methods from Section 7.1. With the Euler method, we established that for a differential equation, $f'(x, y)$, we know from equation 7.1 that,

$$f(x + h) \approx f(x) + hf'(x, y). \qquad\qquad (7.12)$$

We can extend this concept to allow for more than one equation. If we assume there are n differential equations, such that $f_1'(x, Y)$, $f_2'(x, Y)$, $\ldots$, $f_n'(x, Y)$, where $\mathbf{Y}$ is a vector of y-values, for each function, then,

$$f_1(x + h) \approx f_1(x) + hf_1'(x, Y) \qquad\qquad (7.13)$$
$$f_2(x + h) \approx f_2(x) + hf_2'(x, Y) \qquad\qquad (7.14)$$
$$\vdots$$
$$f_n(x + h) \approx f_n(x) + hf_n'(x, Y). \qquad\qquad (7.15)$$

In these equations, the function $f_i(x, Y)$, may depend upon any of the y-values associated with any of the equations. This is functionally identical to the Euler method except for accounting for a vector of y values and the associated function array.

This logic is implemented in Function 7.5, the `eulersys` method. The method places some expectations on the behavior of `f`, the function for which the initial value problem is solved. In particular, it is expected to return an array of *named* objects for the return values of y. If so, it is a trivial matter for `eulersys` to convert the results into a data frame for eventual return to the user. As part of this, the initial values of y are also expected to be named with the corresponding names. This makes using the function more complicated, but the implementation itself is a simple extension of the `euler` method.

```
eulersys <- function(f, x0, y0, h, n) {
    x <- x0
    y <- y0

    ## If y0 values are named, the data frame names them!
    ## The value names produced by f(x, y) should match.
    values <- data.frame(x = x, t(y0))
    for(i in 1:n) {
        y0 <- y0 + h * f(x0, y0)
        x0 <- x0 + h
        values <- rbind(values, data.frame(x = x0, t(y0)))
    }

    return(values)
}
```

R Function 7.5
Euler method for systems of ordinary differential equations

To show **eulersys** in action, we will begin with the example,

$$f_1'(x, Y) = x \qquad (7.16)$$
$$f_2'(x, Y) = y_2 - x \qquad (7.17)$$
$$f_3'(x, Y) = y_3 + x, \qquad (7.18)$$

where y_i is the y value of the ith function, rather than the ith iteration of the differential equation, and Y is a vector of y values. We can begin by implementing this system of differential equations as a single R function:

```
> odesys <- function(x, y) {
+       y1 <- x
+       y2 <- y[2] - y[1]
+       y3 <- y[3] + y[1]
+
+       return(c(y1 = y1, y2 = y2, y3 = y3))
+ }
> odesys(1, c(1, 2, 3))
y1 y2 y3
 1  1  4
```

As we can see, the function, **odesys**, accepts a value of x, Y vector, and returns a new Y vector.

Using this system of equations, we can test it with **eulersys** over $[0, 1]$ where $x = 1$, and $Y = [1, 2, 3]$.

```
> Y <- c(y1 = 1, y2 = 2, y3 = 3)
> eulersys(odesys, 0, Y, 1/2, 2)
```

```
     x    y1    y2 y3
1 0.0 1.00 2.00   3
2 0.5 1.00 2.50   5
3 1.0 1.25 3.25   8
```

As each component is treated identically to the single-equation version of the Euler method, each component of $\mathbf{Y}$ will converge to its correct answer as h decreases, and n increases:

```
> eulersys(odesys, 0, Y, 1/5, 5)
     x    y1   y2      y3
1 0.0 1.00 2.00 3.0000
2 0.2 1.00 2.20 3.8000
3 0.4 1.04 2.44 4.7600
4 0.6 1.12 2.72 5.9200
5 0.8 1.24 3.04 7.3280
6 1.0 1.40 3.40 9.0416
> tail(eulersys(odesys, 0, Y, 1/10, 10), 1)
    x   y1   y2       y3
11 1 1.45 3.45 9.518712
> tail(eulersys(odesys, 0, Y, 1/100, 100), 1)
      x    y1    y2       y3
101 1 1.495 3.495 10.02907
> tail(eulersys(odesys, 0, Y, 1/1000, 1000), 1)
       x     y1     y2       y3
1001 1 1.4995 3.4995 10.08512
```

In fact, each component of $\mathbf{Y}$ has local truncation error of order h^2 and global truncation error of order h.

However, the implementation requires more function evaluations. For each time the function odesys is called, there are three function evaluations internal to that call. These are hidden from the user, as the function odesys will only be called n times. However, for a system of m ordinary differential equations, there are $m \times n$ function evaluations when using eulersys to solve the system's initial value problem.

We can also create higher-order Runge–Kutta methods to solve ordinary differential equations. The implementation is complicated by the record-keeping necessary to ensure that the function's intermediate calculations are handled correctly. However, a fourth-order Runge–Kutta implementation of a system of ordinary differential equations is not more mathematically difficult than the Euler method for a system.

Included with the pracma package is an implementation of the fourth-order Runge–Kutta method. Like rk4, rk4sys includes a different calling structure than we developed with eulersys. Instead of specifying the initial starting point, the step size, and number of steps, like eulersys, the rk4sys function accepts the endpoints along with the number of steps. However, the function produces a good result with $h = 1/10$:

```
> rk4sys(odesys, 0, 1, Y, 10)
$x
 [1] 0.0 0.1 0.2 0.3 0.4 0.5 0.6 0.7 0.8 0.9 1.0

$y
       y1    y2          y3
y0 1.000 2.000   3.000000
   1.005 2.105   3.420854
   1.020 2.220   3.887013
   1.045 2.345   4.404292
   1.080 2.480   4.979121
   1.125 2.625   5.618603
   1.180 2.780   6.330590
   1.245 2.945   7.123758
   1.320 3.120   8.007698
   1.405 3.305   8.993007
   1.500 3.500  10.091399
```

In addition to the different input requirements, rk4sys returns its results as a list including both x values as a vector and a data frame containing the y values.

7.2.2 Boundary Value Problems

Using equations 7.10 and 7.11 with a method to solve an initial value problem is only part of the solution of a boundary value problem. In the boundary value problem, we are given $f''(x, y)$, $f(x_0) = y_0$, and $f(x_n) = y_n$. However, when we solve a system of ordinary differential equations for equations 7.10 and 7.11, the result is an estimate of $f(x_n)$, which is not an input and something we already know. Further, we must provide a value of $f'(x_0)$ to the ordinary differential equation solver, the value we are searching for.

We address this problem by treating it as a root-finding exercise. We can create a function in R that accepts as an argument a value $f'(x_0)$ and returns the difference between the known value of y_n and estimated value found. The closer that value is to 0, the closer our guess at $f'(x_0)$ is. This is called the shooting method, and is not implemented here directly, due to complexity of the necessary embedded functions.

However, for example, we can image a boundary value problem such that, $f''(x, y) = y^2 - 2$, $f(0) = 1$, and $f(1) = 1$. This function is implemented in Function 7.6, which uses eulersys on an embedded system of ordinary differential equations. Finally, it subtracts the known value of y_n and returns the difference. If we plot the values of bvpexample for different values of $f'(x)$, shown in Figure 7.7, we can see the value of bvpexample is evidentially negative when $f'(x) = -2$ and positive when $f'(x) = 2$.

```
bvpexample <- function(x) {
    x0 <- 0
    y0 <- c(y1 = 1, y2 = x)
    yn <- 1

    odesystem <- function(x, y) {
        y1 <- y[2]
        y2 <- y[1]^2 - 2

        return(c(y1 = y1, y2 = y2))
    }

    z <- eulersys(odesystem, x0, y0, 1 / 1000, 1000)
    tail(z$y1, 1) - yn
}
```

R Function 7.6
Example boundary value problem, $f''(x, y) = y^2 - 2$, $f(0) = 1$, and $f(1) = 1$

Further, we can estimate the value of bvpexample at specific discrete values of $f'(x)$.

```
> bvpexample(-2)
[1] -2.782773
> bvpexample(-1)
[1] -1.775453
> bvpexample(0)
[1] -0.5779241
> bvpexample(1)
[1] 0.8409145
> bvpexample(2)
[1] 2.518243
```

This is a problem we can then solve with our family of root-finding algorithms. For instance, neither bisection nor secant require the calculation of the first derivative of bvpexample and both functions can be used to estimate the value of $f'(x_0)$.

```
> (bvp.b <- bisection(bvpexample, 0, 1))
[1] 0.4282227
> (bvp.s <- secant(bvpexample, 0))
[1] 0.4278309
```

We can verify our correctness of these estimates using the Euler method.

```
> odesystem <- function(x, y) {
+     y1 <- y[2]
+     y2 <- y[1]^2 - 2
```

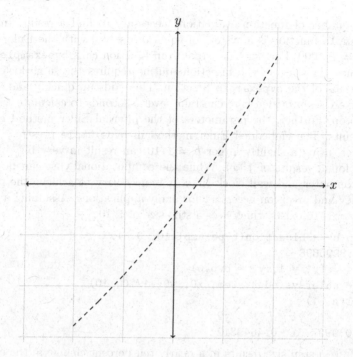

Figure 7.7
Plot of bvpexample for $x \in [-2, 2]$

```
+
+        return(c(y1 = y1, y2 = y2))
+ }
> x0 <- 0
> y0 <- c(y1 = 1, y2 = bvp.b)
> yn <- 1
> z <- eulersys(odesystem, x0, y0, 1/1000, 1000)
> tail(z, 1)
       x         y1          y2
1001 1 1.000548 -0.4276659
> y0 <- c(y1 = 1, y2 = bvp.s)
> z <- eulersys(odesystem, x0, y0, 1/1000, 1000)
> tail(z, 1)
       x         y1          y2
1001 1 0.999999 -0.4285557
```

Whether either result is sufficiently correct, like other numerical problems, is subject to the requirements of the application.

The downside to the shooting method for solving boundary value problems is the number of function evaluations necessary to find a result. Inside the bvpexample function is a system of equations solved with the Euler method and $h = 1/1000$. Further, the bisection function calls bvpexample numerous times. This use of the bisection routine requires the same $\log_2 \frac{b-a}{t} - 1$ evaluations of the bvpexample function. This adds up quickly and even the basic exercises provided here can take several seconds to calculate on a modest laptop. Further, the parameters of the internal Euler method can alter the results. The embedded Euler method in bvpexample uses $h = 1/1000$. However, using a slightly larger $h = 1/10$ can result in very big differences in the found results of $f'(x)$. While the computational cost can be high, in both computer and wall clock time, it is necessary to achieve the degree of accuracy and precision necessary for many applications. Assuming a function bvpexample10 exists which uses a step size of $1/10$,

```
> (bvp.b <- bisection(bvpexample10, 0, 1))
[1] 0.3920898
> y0 <- c(y1 = 1, y2 = bvp.b)
> z <- eulersys(odesystem, x0, y0, 1/10, 10)
> tail(z, 1)
     x          y1          y2
11 1 0.9997761 -0.4649312
```

The shift in step size results in a nearly ten percent change in the estimated value of $f'(x)$. Using a smaller step size to create a plot of the boundary value problem, similar to Figure 7.7, will allow us to narrow down the search space before using a root-finding algorithm. This would reduce the overall number of function evaluations necessary to solve boundary value problems.

7.3   Partial Differential Equations

Beyond ordinary differential equations are partial differential equations, any equation that contains a partial derivative. These equations can be solved numerically, which is especially valuable when an analytic solution is difficult. In this section, instead of developing general solutions to problems, we will develop specific solutions to two broad problems, the heat equation and the wave equation.

7.3.1   The Heat Equation

The heat equation is a classic example of a partial differential equation. The equation itself describes how heat dissipates across a uniform object. For instance, imagine heat, such as from a soldering iron, were applied to a sheet of metal. After the iron were removed, the point where heat was applied will be hot while the edges could be very cool. Over several minutes, the sheet of metal will get cooler where the heat was applied, but warms up overall as the heat transfers through the sheet. The heat equation models this change in temperature. In one dimension, the heat equation describes the change in temperature of a thin rod or wire and in one dimension, the heat equation is,

$$\frac{\partial u}{\partial t} = \alpha \frac{\partial^2 u}{\partial x^2}, \tag{7.19}$$

where α is the thermal diffusivity, a ratio describing how quickly a substance cools down after heating. Though different from an ordinary differential equation, an approach conceptually similar to the Euler method can be used to model the heat equation. This approach is the family of finite difference methods. One particular method, called forward-time central-space (FTCS), is well suited to solving the heat equation. Starting with equation 7.19, we know we can model the right-hand side as a single function $F()$, such that,

$$\frac{\partial u}{\partial t} = F(u, x, t, \frac{\partial^2 u}{\partial x^2}). \tag{7.20}$$

In practice, the embedded partial differential in $F()$ requires α, Δx, and Δt, so, allowing that u is a vector of observations separated by Δx,

$$\frac{u_{i+1} - u_i}{\Delta t} \approx F(u, \alpha, x, \Delta x, t, \Delta t). \tag{7.21}$$

Solving equation 7.21 for u_{i+1} yields,

$$u_{i+1} \approx u_i + \alpha \frac{\Delta t}{\Delta x^2} F(u, x, t). \tag{7.22}$$

At the level of the elements of u,

$$u_{i+1,j} \approx u_{i,j} + \alpha \frac{\Delta t}{\Delta x^2} (u_{i,j-1} - 2u_{i,j} + u_{i,j+1}), \tag{7.23}$$

where u_{ij} is such that i is the iteration and j is the element of u. Equation 7.23 completely describes the process for modeling the heat equation and is presented in Function 7.7, the **heat** function.

The function itself actually performs two different sets of value productions. First, in space, or along u, the method converges with second-order efficiency, due to the use of both leftward and rightward inclusion in the calculation, in addition to the last value of the element. Second, the method is first order in time, from one iteration of u to the next, as it only looks back one time step. This efficiency though is sufficient for modelling heat dissipation.

We start with an initial u concentrated on one end, but with both endpoints at 0. We will use the sin function to create this initial pattern with a quartic function to enforce the concentration. The function will be iterated over a brief 25 steps with $\Delta t = 0.001$ and 21 points along the x-axis, including the endpoints.

```
heat <- function(u, alpha, xdelta, tdelta, n) {
    m <- length(u)
    uarray <- matrix(u, nrow = 1)
    newu <- u

    h <- alpha * tdelta / xdelta^2
    for(i in 1:n) {
        for(j in 2:(m - 1)) {
            ustep <- (u[j - 1] + u[j + 1] - 2 * u[j])
            newu[j] <- u[j] + h * ustep
        }
        u <- newu
        u[1] <- u[m]
        uarray <- rbind(uarray, u)
    }

    return(uarray)
}
```

R Function 7.7
The heat equation via the forward-time central-space method

```
> alpha <- 1
> x0 <- 0
> xdelta <- .05
> x <- seq(x0, 1, xdelta)
> u <- sin(x^4 * pi)
> tdelta <- .001
> n <- 25
> z <- heat(u, alpha, xdelta, tdelta, n)
```

The results are captured in z and we will use the t and head functions to display just the first two rows.

```
> t(head(z, n = 2))
                              u
 [1,] 0.000000e+00 1.224647e-16
 [2,] 1.963495e-05 1.295907e-04
 [3,] 3.141593e-04 7.068581e-04
 [4,] 1.590431e-03 2.454361e-03
 [5,] 5.026527e-03 6.550093e-03
 [6,] 1.227154e-02 1.464258e-02
 [7,] 2.544415e-02 2.884787e-02
 [8,] 4.712606e-02 5.173811e-02
 [9,] 8.033810e-02 8.630561e-02
[10,] 1.284689e-01 1.358651e-01
```

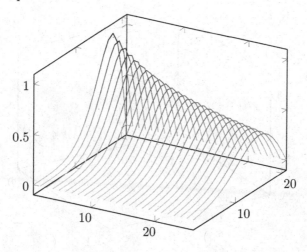

Figure 7.8
Heat equation plot for $\Delta t = 0.001$

```
[11,]  1.950903e-01  2.038185e-01
[12,]  2.835321e-01  2.931403e-01
[13,]  3.959943e-01  4.053552e-01
[14,]  5.318587e-01  5.386799e-01
[15,]  6.847761e-01  6.849886e-01
[16,]  8.382247e-01  8.255323e-01
[17,]  9.599423e-01  9.263228e-01
[18,]  9.976112e-01  9.363565e-01
[19,]  8.821433e-01  7.956006e-01
[20,]  5.503186e-01  4.629210e-01
[21,]  1.224647e-16  1.224647e-16
```

The results are better understood if displayed as a three-dimensional plot where time is one dimension. That plot is shown in Figure 7.8.

As remarkably as this implementation works, we can also see these results go wrong quickly. The heat equation with FTCS is stable provided,

$$\alpha \frac{\Delta t}{\Delta x^2} \leq \frac{1}{2} \tag{7.24}$$

In the example shown, $1 \times 0.001/0.05^2 = 0.4$. Changing $\Delta t = 0.0015$ leads to a value of 0.6 and we can see in Figure 7.9 the result starts the same way, but after a mere 25 steps, has already gone haywire producing wild jumps and negative values.

In general, this cannot be run out to the long term, or even any longer than it has been run here, while still providing a reliable result. This can be addressed by using a smaller time step, but a smaller time step requires more

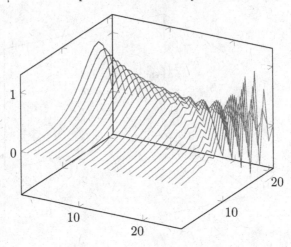

Figure 7.9
Heat equation plot for $\Delta t = 0.0015$

iterations to reach t_n, the required ending time step, or using a larger Δx, the distance between observation points along u. This reduces the level of detail available in the final result. One tradeoff or the other is necessary, depending on the requirements of the analysis.

　　The heat equation can be extended to more than one dimension. More remarkably, the heat equation is a generalized form of many different problems from wildly different fields. Among others, the Black–Scholes option pricing model, used in finance, can be reduced to the heat equation. Partial differential equations representing particle diffusion in particle physics and trait diffusion in genetics are also well modeled using the heat equation.

7.3.2　The Wave Equation

The wave equation is another classic example of a partial differential equation. It comes in several variants and has applications beyond the name. In principle, the wave equation describes the path of a wave traveling through a medium. For a one-dimensional wave equation, this describes a wave traveling on a string, like a violin's string. In two dimensions, the wave equation describes a wave on a membrane, like a drumhead. And for three dimensions, it describes the propagation of sound through the air. The equation can also describe light waves.

　　For our purposes, we will look at a one-dimensional wave using a second-order differential equation, like a wave on a string. And to further simplify the problem, we will assume the string being modeled is fixed at both ends. This

```
wave <- function(u, alpha, xdelta, tdelta, n) {
    m <- length(u)
    uarray <- matrix(u, nrow = 1)
    newu <- u

    h <- ((alpha * tdelta) / (xdelta))^2

    ## Initial the zeroth timestep
    oldu <- rep(0, m)
    oldu[2:(m - 1)] <- u[2:(m - 1)] + h *
        (u[1:(m - 2)] - 2 * u[2:(m - 1)] + u[3:m]) / 2

    ## Now iterate
    for(i in 1:n) {
        ustep1 <- (2 * u - oldu)
        ustep2 <- u[1:(m - 2)] - 2 * u[2:(m - 1)] + u[3:m]
        newu <- ustep1 + h * c(0, ustep2, 0)
        oldu <- u
        u <- newu
        uarray <- rbind(uarray, u)
    }

    return(uarray)
}
```

R Function 7.8.
The wave equation via finite differences

is the same circumstance as a string on a musical instrument, or a jumprope with the ends held in place, after it has been plucked.

In one dimension, the wave equation is similar to the heat equation,

$$\frac{\partial^2 u}{\partial t^2} = \alpha^2 \frac{\partial^2 u}{\partial x^2}, \tag{7.25}$$

where α is the speed of the wave. Different texts use a variety of letters including c and γ to represent the wave speed. We use α to show the similarities to the heat equation as both equations effectively show transport through a medium, though of different kinds.

On the other hand, there are also differences. This form of the wave equation is a second-order partial differential equation, requiring a slightly different approach from a finite differences perspective. We need to approximate both sides of the equation. Using the same approach as with the heat equation,

$$\frac{\partial^2 u}{\partial x^2} \approx \frac{u_{i-1,j} - 2u_{i,j} + u_{i+1,j}}{\Delta x^2}. \tag{7.26}$$

governs the wave equation in space. In time, the wave equation is governed by the similar,

$$\frac{\partial^2 u}{\partial t^2} \approx \frac{u_{i,j-1} - 2u_{i,j} + u_{i,j+1}}{\Delta t^2}, \tag{7.27}$$

In both equations 7.27 and 7.26, i is the ith member in space and j represents the time step.

Together, we can substitute these back into equation 7.25 yielding, a final form of the equation,

$$u_{i,j+1} = 2u_{i,j} - u_{i,j-1} + \left(\frac{\alpha \Delta x}{\Delta t}\right)^2 (u_{i-1,j} - 2u_{i,j} + u_{i+1,j}), \tag{7.28}$$

emerges, as the definitive final form of the approximation. However, the heat equation uses only the prior step's information to calculate the current iteration. This implementation of finite differences uses two prior steps, and we only have one as we enter the function. Some implementations of the wave equation address this by accepting two starting conditions, one at t_0 and one at t_1, and subsequent steps are modeled from these two starting points.

Other implementations take a different approach and model the wave from a single initial condition and simulating the second initial condition. In the most naïve methods, the initial and second conditions are equal. We take a slightly more advanced approach and model the second condition using an approach similar to the heat equation. This echoes the approach used when creating two-step linear methods as in Section 7.1.3, where we created a step using another method and began our multistep process. This process is implemented in `wave`, given in Function 7.8.

We take an approach here that instead of generating a first second step using the simpler method, we actually take a step backwards and create a "zeroth" step, as shown in the function. This helps with bookkeeping; rather than forcing a recording of both the first and second steps, before entering the loop, we record the first and continue into the loop for n steps. Inside the loop, the step is broken down into three distinct parts and the next u iteration is assembled from those parts. Then it is recorded in the output array and both the previous and current iterations are incremented.

We can see this at work in the following example. Of interest, we set the initial state in u to be a partial sine wave. The left side, the lower half of the array, is the positive half of a sine wave. The right half is zeroed out, leading to a wave that travels down the string. The process fixes the ends of the string at zero, making them unmoving. The wave will reach the end of the line and bounce back to the start.

```
> speed <- 2
> x0 <- 0
> xdelta <- .05
> x <- seq(x0, 1, xdelta)
> m <- length(x)
> u <- sin(x * pi * 2)
```

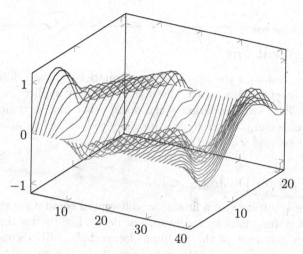

Figure 7.10
Wave equation plot for $\Delta t = 0.001$

```
> u[11:21] <- 0
> tdelta <- .02
> n <- 40
> z <- wave(u, speed, xdelta, tdelta, n)
```

We can capture the results, but they are better plotted. Shown in Figure 7.10, the string is shown at several timesteps in succession. The wave is shown travelling down the string to the end and starting its return.

This method remains stable provided

$$\left(\alpha \frac{\Delta t}{\Delta x^2}\right)^2 \leq \frac{1}{2}. \tag{7.29}$$

We can create unstable versions very quickly by altering the number of points evaluated, the timestep size, or the speed of the wave. When the value just exceeds the thresholds, the analysis will remain stable for a handful of iterations before the results travel out of reason. Those results resemble a string snapping while vibrating.

The wave equation can extend to an infinite number of dimensions, though more than a few stop modeling reality. Further, the applications come up in modeling earthquakes, general relativity, and sound traveling through air.

7.4 Applications

As we have seen with the heat and wave equations, partial differential equations have direct applications in physics and ancillary applications across other fields. But differential equations turn up in other places, typically any field where calculus can play a role. We are going to look at two more examples, a decay function and a competition question.

7.4.1 Carbon Dating

One simple application of a first-order differential equation is the decay function. A decay function can model nuclear decay, like the breakdown of radon gas. Radon gas, part of the uranium decay chain, often seeps into ground floors and basements where trace uranium is in the ground. Radon decays into polonium, itself a radioactive and very dangerous substance.

Another example of radioactive decay is carbon-14. Carbon typically comes in an isotope called carbon-12, which is very stable. Another isotope that appears in nature is carbon-14. Carbon-14 has a half-life of 5730 years, meaning that every 5730 years, approximately half of the carbon-14 would have decayed into nitrogen-14, a stable isotope. Carbon-14 is constantly being generated through the effects of solar and space radiation in the upper atmosphere and living organisms absorb this over time, a process that stops when the organism dies. By measuring the amount of carbon-14 left in an organism, we can estimate when something died, a process commonly known as carbon dating.

Decay functions are generally given as,

$$f(t) = x_0 e^{kt}, \tag{7.30}$$

where x_0 is some initial value, k is the decay rate, and t is the amount of time. The associated differential equation is,

$$\frac{dx}{dt} = kx. \tag{7.31}$$

If k is negative, this is a decay function. If k were positive, the function would represent exponential growth, but we are more interested in the decay function right now. Finding k, however, is important for this process to work. Knowing that after 5730 years, one half of the carbon-14 will be left, and assuming that we started with 100 percent of the carbon-14 in place, we can solve for k:

$$1/2 = f(5730)/f(0) \tag{7.32}$$

$$= e^{5730k}/e^{0k} \tag{7.33}$$

$$= e^{5730k} \tag{7.34}$$

$$\ln\frac{1}{2} = 5730k \tag{7.35}$$

$$\frac{-\ln 2}{5730} = k \tag{7.36}$$

Therefore, $k \approx -0.0001209681$, which is a very small number, but the decay rate per time period, and we are measuring t in years. Using this, developing a carbon-dating function is manageable. In accordance with how our Runge–Kutta methods are developed, we will use y to represent the amount of carbon-14, as a percentage remaining, and x to represent the amount of time passed. Therefore, the initial value of $x = 0$ and $y = 1$.

```
> carbondating <- function(x, y) { y * -log(2) / 5730 }
> years = 3000
> steps = 10        # Per year
> tail(rungekutta4(carbondating, 0, 1, 1 / steps, steps * years))
            x       y
29996 2999.5 0.6956951
29997 2999.6 0.6956867
29998 2999.7 0.6956783
29999 2999.8 0.6956698
30000 2999.9 0.6956614
30001 3000.0 0.6956530
```

It takes several seconds on a modest laptop, but there are last few steps of the fourth-order Runge–Kutta process. We can execute a small sanity check on this, remembering that after 5730 years, we can expect exactly one half of the carbon-14 to remain:

```
> years = 5730
> steps = 10        # Per year
> tail(rungekutta4(carbondating, 0, 1, 1 / steps, steps * years))
            x       y
57296 5729.5 0.5000302
57297 5729.6 0.5000242
57298 5729.7 0.5000181
57299 5729.8 0.5000121
57300 5729.9 0.5000060
57301 5730.0 0.5000000
```

A result correct within display precision. But even the first-order Euler method, with one time step per year is very close:

```
> years = 5730
> steps = 1      # Per year
> tail(euler(carbondating, 0, 1, 1 / steps, steps * years))
          x        y
5726 5725 0.5002816
5727 5726 0.5002210
5728 5727 0.5001605
5729 5728 0.5001000
5730 5729 0.5000395
5731 5730 0.4999790
```

A result that is sufficiently close for many purposes.

This sort of function works for any decay process. Other atomic decay systems can be modeled simply by changing the value of k appropriately. Beer froth follows an exponential decay model (Leike 2002) and decay models have a variety of applications in finance and economics.

7.4.2   Lotka–Volterra Equations

Another application of differential equations come from the Lotka–Volterra equations (Murray 2002, 79–86). The equations, also known as the predator-prey equations, model the relationship between two species when one is a primary food source for the other. In this simplified model, we can predict the rise and fall of the food source and how it affects the predator. The Lotka–Volterra equations are,

$$\frac{dx}{dt} = \alpha x - \beta xy \tag{7.37}$$

$$\frac{dy}{dt} = \delta xy - \gamma y \tag{7.38}$$

where x is the population of the food source, y is the population of the prey, and α, β, γ, and δ describe the relationship between the predator and prey, and t is time.

Values for α, β, γ, and δ can be estimated from population studies. The variable α describes the natural growth rate for prey and β is the rate at which the predator kills prey. Similarly, the variable γ is the natural death rate for predators and δ is a growth rate for predators, predicated on a reliable food source, the prey. As we can see, the two populations are interlinked and growth in prey will lead to growth in predators, until the predators overconsume and prey population declines.

For our purposes, we will use values for hypothetical, but visually interesting, growth and death rates so that,

$$\alpha = 0.8 \tag{7.39}$$
$$\beta = 0.1 \tag{7.40}$$
$$\gamma = 0.2 \tag{7.41}$$
$$\delta = 0.01 \tag{7.42}$$

and use `eulersys` to carry out the model with an initial population of 20 prey and 10 predators in a closed system, like an island.

```
> lotkavolterra <- function(x, y) {
+     alpha <- 0.8
+     beta <- 0.1
+     gamma <- 0.2
+     delta <- 0.01
+
+     y1 <- alpha * y[1] - beta * y[1] * y[2]
+     y2 <- delta * y[1] * y[2] - gamma * y[2]
+     return(c(y1 = y1, y2 = y2))
+ }
> init <- c(y1 = 20, y2 = 10)
> pop <- eulersys(lotkavolterra, 0, init, 1/10, 50 * 10)
```

The new population figures are stored in *pop* and are best shown through two plots over time. Figure 7.11 shows the top line is dotted and represents prey over time, t. The bottom solid line shows the population of predators. There is a recurring spike in the population of prey followed by a spike in predators several time steps later.

Important in this model is that the spike in both populations gets larger with each cycle. This is an artifact of the Euler method and is purely numerical error. If this same process were modeled using a fourth-order Runge–Kutta process, such as with `rk4sys` in the `pracma` package, the spike height would be much more stable over successive cycles.

Variations on the Lotka–Volterra have applications outside of biological population dynamics. Economists have used the equations to model competitive markets, to estimate the interaction between different businesses by size and output. The equations can also scale beyond the simple predator-prey arrangement and encompass more complex population relationships, including numerous species.

Comments

Differential equations encompass so many problems, strategies, and methods that solving differential equations is its own subfield of numerical analysis.

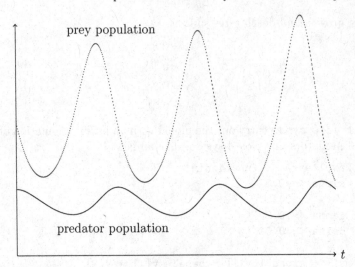

Figure 7.11
Lotka–Volterra model for a hypothetical population dynamic

This chapter has included only a high-level overview of the problems and elementary solutions to some of the most common and well-understood methods. Variants on the finite differences method given in this chapter are the strongest method for most interesting differential equations problems. However, for an ordinary differential equation that is not particularly interesting, the Runge–Kutta methods are sufficient for solving the problem.

The problem of partial differential equations is conceptually similar, but solved differently from ordinary differential equations. There is a wealth of methods for solving partial differential equations, as for everything else in this book. However, many interesting partial differential equations are mathematically equivalent to either the heat or wave equations, as discussed. Accordingly, the methods for solving each can adapt fairly easily to other problems.

In addition to those functions mentioned, the `pracma` package for R provides basic numerical methods for linear algebra, integration, and differential equations. We have seen the `rk4` function in this package can be used to quickly and efficiently solve ordinary differential equations, though the results are not significantly different from our own implementation of the fourth-order Runge–Kutta method.

Beyond `pracma`, there is a specialized package called `deSolve` that focuses on solving differential equations. Specialized and high-speed methods for solving initial value problems, boundary value problems, plotting, and other differential equation problems are given. The `deSolve` package also provides optimized versions of its methods for handling one-, two-, and three-dimensional differential equations.

The `bvpSolve` package provides specific functions for solving boundary value problems, including another implementation of the shooting method. Finally, the `deSolve` and `ReacTran` packages can also solve ordinary differential equations that are the result of converting partial differential equations. The `rootSolve` package can solve time-invariant partial differential equations.

Again, this broad set of packages for different classes of differential equations reflects the broad classification of differential equations. Finding the right package for a problem should not be difficult, provided we understand both the problem and its application. The CRAN repository provides a differential equations task view containing a number of packages for solving general and specific differential equations.

Exercises

1. Show the first 10 steps of the Euler method for the differential equation $f'(x,y) = y$ where $x_0 = 0$ and $y_0 = 2$ and step size of $h = 0.1$.

2. Show the first 10 steps of the fourth-order Runge–Kutta method for the differential equation $f'(x,y) = y$ where $x_0 = 0$ and $y_0 = 2$ and step size of $h = 0.1$.

3. In the example $f'(x,y) = -10y$, shown in Figure 7.3, the results are shown to be sensitive to step-size. Sketch the vector field for the differential equation at $h = 1/2$, $h = 1/5$, and $h = 1/10$.

4. Show the first 10 steps of the Adams–Bashforth method for the differential equation $f'(x,y) = y$ where $x_0 = 0$ and $y_0 = 2$ and step size $h = 0.2$.

5. Using equation 7.30, with $k = 2$, show whether or not function `rungekutta4` is A-stable.

6. Using equation 7.30, with $k = 11$, show whether or not function `euler` is A-stable.

7. There is a second solution to the boundary value problem presented in Section 7.2.2. Find it.

8. Using the information from Section 7.4.1, a decaying organism has 20 percent of its original carbon-14 present. How long ago did it die?

9. Radon-222 has a half-life of 3.8 days. Assuming you start with 400 grams, using any Runge–Kutta method, after 10 days, how much radon-222 will remain?

10. Implement a third-order Adams–Bashforth method, given in equation 7.9.

Suggested Reading

Abramowitz, Milton, and Irene A Stegun. 1964. *Handbook of Mathematical Functions: with Formulas, Graphs, and Mathematical Tables*. 55. Courier Corporation.

Adler, Joseph. 2010. *R in a Nutshell: A Desktop Quick Reference*. Sebastopol, California: O'Reilly Media.

Atkinson, Kendall E. 1989. *An Introduction to Numerical Analysis*. 2nd ed. Chichester, New York: John Wiley & Sons.

Bandelow, Christoph. 1982. *Inside Rubiks Cube and Beyond*. Boston: Birkhauser.

Benvenuti, Cristoforo, and P Chiggiato. 1993. "Obtention of Pressures in the 10^{-14} Torr Range by Means of a Zr-V-Fe Non Evaporable Getter." *Vacuum* 44 (5): 511–513.

Bertsimas, Dimitris, and John Tsitsiklis. 1993. "Simulated Annealing." *Statistical Science* 8 (1): 10–15.

Björck, Åke. 2015. *Numerical Methods in Matrix Computations*. Springer.

Bloomfield, Victor A. 2014. *Using R for Numerical Analysis in Science and Engineering*. The R Series. Boca Raton, Florida: Chapman and Hall/CRC.

Bojanov, Borislav D, H Hakopian, and B Sahakian. 2013. *Spline Functions and Multivariate Interpolations*. Vol. 248. Mathematics and Its Applications. London: Springer.

Bonnans, Joseph-Frédéric, et al. 2003. *Numerical Optimization: Theoretical and Practical Aspects*. Berlin: Springer.

Buzzi-Ferraris, Guido, and Flavio Manenti. 2010. *Interpolation and Regression Models for the Chemical Engineer: Solving Numerical Problems*. John Wiley & Sons.

Byrne, Charles L. 2008. *Applied Iterative Methods*. Wellesley, Massachusetts: AK Peters.

Cormen, Thomas H, et al. 2009. *Introduction to Algorithms*. Cambridge, Massachusetts: MIT Press.

Dalgaard, Peter. 2002. *Introductory Statistics with R*. New York: Springer.

Davies, Paul. 2008. *The Goldilocks Enigma: Why is the Universe Just Right for Life?* Boston: Houghton Mifflin Harcourt.

Davis, Philip J. 2014. *Interpolation and Approximation*. Mineola, New York: Dover Publications.

Davis, Philip J, and Philip Rabinowitz. 2007. *Methods of Numerical Integration*. 2nd ed. Mineola, New York: Dover Publications.

Deuflhard, Peter, and Andreas Hohmann. 2003. *Numerical Analysis in Modern Scientific Computing: An Introduction*. Vol. 43. Texts in Applied Mathematics. New York: Springer.

Epperson, James F. 2014. *An Introduction to Numerical Methods and Analysis*. 2nd ed. Somerset, New Jersey: John Wiley & Sons.

Fisher, Ronald A. 1936. "The Use of Multiple Measurements in Taxonomic Problems." *Annals of Eugenics* 7 (2): 179–188.

Friedman, Milton. 1962. "The Interpolation of Time Series by Related Series." *Journal of the American Statistical Association* 57 (300): 729–757.

Garay, Luis J. 1995. "Quantum Gravity and Minimum Length." *International Journal of Modern Physics A* 10 (02): 145–165.

Gerber, Leon. 2007. "A Quintile Rule for the Gini Coefficient." *Mathematics Magazine* 80 (2): 133–135.

Gibb, David. 1915. *A Course in Interpolation and Numerical Integration for the Mathematical Laboratory*. London: G. Bell & Sons.

Greenbaum, Anne, and Timothy P Chartier. 2012. *Numerical Methods: Design, Analysis, and Computer Implementation of Algorithms*. Princeton, New Jersey: Princeton University Press.

Hamming, Richard. 2012. *Numerical Methods for Scientists and Engineers*. Dover Publications.

Henderson, Darrall, Sheldon H Jacobson, and Alan W Johnson. 2003. "The Theory and Practice of Simulated Annealing." In *Handbook of Metaheuristics*, ed. by Fred Glover and Gary A. Kochenberger, 287–319. Boston: Kluwer Academic Publishers.

Himmelblau, David Mautner. 1972. *Applied Nonlinear Programming*. New York: McGraw-Hill Companies.

Isaacson, Eugene, and Herbert Bishop Keller. 2012. *Analysis of Numerical Methods*. Dover Publications.

Jones, Owen, Robert Maillardet, and Andrew Robinson. 2014. *Scientific Programming and Simulation Using R*. 2nd ed. The R Series. Boca Raton, Florida: Chapman and Hall/CRC.

Judd, Kenneth L. 1998. *Numerical Methods in Economics*. Boston: MIT Press.

Kahan, William. November 20, 2004. "On the Cost of Floating-Point Computation Without Extra-Precise Arithmetic." http://www.cs.berkeley.edu/~wkahan/Qdrtcs.pdf.

Kincaid, David Ronald, and Elliott Ward Cheney. 2009. *Numerical Analysis: Mathematics of Scientific Computing.* 3rd ed. Vol. 2. The Sally Series; Pure and Applied Undergraduate Texts. Providence, Rhode Island: American Mathematical Society.

Kirkpatrick, Scott, C Daniel Gelatt, Mario P Vecchi, et al. 1983. "Optimization by Simulated Annealing." *Science* 220 (4598): 671–680.

Knuth, Donald E. 1997. *The Art of Computer Programming: Seminumerical Algorithms.* 2nd ed. Vol. 3. Boston: Addison-Wesley Longman Publishing.

Kutz, J Nathan. 2013. *Data-Driven Modeling & Scientific Computation: Methods for Complex Systems & Big Data.* Oxford: Oxford University Press.

Kythe, Prem K, and Michael R Schäferkotter. 2005. *Handbook of Computational Methods for Integration.* Boca Raton, Florida: Chapman and Hall/CRC.

Lambert, Peter J. 2001. *The Distribution and Redistribution of Income.* 3rd ed. Manchester: Manchester University Press.

Lander, Jared P. 2013. *R for Everyone: Advanced Analytics and Graphics.* Addison-Wesley Data & Analytics. Upper Saddle River, New Jersey: Addison-Wesley Professional.

Leike, Arnd. 2002. "Demonstration of the Exponential Decay Law Using Beer Froth." *European Journal of Physics* 23 (1): 21.

Massopust, Peter. 2010. *Interpolation and Approximation with Splines and Fractals.* New York: Oxford University Press.

Matloff, Norman. 2011. *The Art of R Programming: A Tour of Statistical Software Design.* San Francisco, California: No Starch Press.

Murray, James D. 2002. *Mathematical Biology I: An Introduction.* New York.

Olver, Frank WJ. 2010. *NIST Handbook of Mathematical Functions.* New York: Cambridge University Press.

Phillips, George M. 2003. *Interpolation and Approximation by Polynomials.* Vol. 14. Berlin: Springer.

Plato, Robert. 2003. *Concise Numerical Mathematics.* Trans. by Richard Le Borne and Sabine Le Borne. Graduate Studies in Mathematics 57. Providence, Rhode Island: American Mathematical Society.

Press, William H, et al. 2007. *Numerical Recipes: The Art of Scientific Computing.* 3rd ed. New York: Cambridge University Press.

Schatzman, Michelle. 2002. *Numerical Analysis: A Mathematical Introduction.* Trans. by John Taylor. Oxford: Clarendon Press.

Scott, L Ridgway. 2014. *Numerical Analysis.* Princeton, New Jersey: Princeton University Press.

Sedgewick, Robert, and Kevin Wayne. 2011. *Algorithms.* Vol. 4. Upper Saddle River, New Jersey: Addison-Wesley.

Severance, Charles. 1998. "IEEE 754: An Interview with William Kahan." *Computer* 31 (3): 114–115.

Shampine, Lawrence F, Richard C Allen, and Steven Pruess. 1997. *Fundamentals of Numerical Computing*. New York: Wiley.

Späth, Helmuth. 1995a. *One Dimensional Spline Interpolation Algorithms*. Wellesley, Massachusetts: AK Peters.

— . 1995b. *Two Dimensional Spline Interpolation Algorithms*. Wellesley, Massachusetts: AK Peters.

Steffensen, Johan Frederik. 2006. *Interpolation*. Mineola, New York: Dover Publications.

Szabados, József, and Péter Vértesi. 1990. *Interpolation of Functions*. Singapore: World Scientific.

Wiener, Norbert. 1949. *Extrapolation, Interpolation, and Smoothing of Stationary Time Series*. Vol. 2. Cambridge, Massachusetts: MIT Press.

Wilkinson, JH. 1959. "The Evaluation of the Zeros of Ill-Conditioned Polynomials." *Numerische Mathematik* 1 (1): 150–166.

Yorke, James A, and William N Anderson Jr. 1973. "Predator-Prey Patterns." *Proceedings of the National Academy of Sciences of the United States of America* 70 (7): 2069.

Zuras, Dan, et al. 2008. "IEEE Standard for Floating-Point Arithmetic." *IEEE Std 754-2008*: 1–70.

Zuur, Alain, Elena N. Ieno, and Erik Meesters. 2009. *A Beginner's Guide to R*. Use R! New York: Springer.

Index

Printed in the United States
by Baker & Taylor Publisher Services

Printed in the United States
by Baker & Taylor Publisher Services